November 7, 2013

Dear Members of the Aviation Community:

 I am pleased to present the Federal Aviation Administration's (FAA) Roadmap for *Integration of Civil Unmanned Aircraft Systems (UAS) in the National Airspace System (NAS)*. The FAA and the UAS Aviation Rulemaking Committee (ARC) worked together for the past year to produce this roadmap. Unmanned aircraft offer new ways for commercial enterprises and public operators to increase operational efficiency, decrease costs, and enhance safety; and this roadmap will allow us to safely and efficiently integrate them into the NAS.

The FAA is committed to the safe and efficient integration of UAS into the NAS. However, as safety is our top priority, UAS integration must be accomplished without reducing existing capacity, decreasing safety, impacting current operators, or placing other airspace users or persons and property on the ground at increased risk. We have made great progress in accommodating public UAS operations, but challenges remain for the safe, long-term integration of both public and civil UAS in the NAS.

This roadmap outlines the actions and considerations needed to enable UAS integration into the NAS. The roadmap also aligns proposed FAA actions with Congressional mandates from the *FAA Modernization and Reform Act of 2012*. This plan also provides goals, metrics, and target dates for the FAA and its government and industry partners to use in planning key activities for UAS integration.

We will update the specific implementation details (goals, metrics, target dates) as we learn from our current UAS operations, leverage ongoing research, and incorporate the work of our government and industry partners in all related areas.

Thank you for your continued support and active participation in the safe and efficient integration of UAS in the NAS.

Michael P. Huerta
Administrator

Table of Contents

Executive Summary

Expanding Operations of Unmanned Aircraft Systems in the NAS

Since the early 1990s, unmanned aircraft systems (UAS) have operated on a limited basis in the National Airspace System (NAS). Until recently, UAS mainly supported public operations, such as military and border security operations. The list of potential uses is now rapidly expanding to encompass a broad range of other activities, including aerial photography, surveying land and crops, communications and broadcast, monitoring forest fires and environmental conditions, and protecting critical infrastructures. UAS provide new ways for commercial enterprises (civil operations) and public operators to enhance some of our nation's aviation operations through increased operational efficiency and decreased costs, while maintaining the safety of the NAS.

As stated in *Destination 2025* (2011):

> *"The Federal Aviation Administration's (FAA) mission is to provide the safest, most efficient aviation system in the world. What sets the United States apart is the size and complexity of our infrastructure, the diversity of our user groups, our commitment to safety and excellence, and a history of innovation and leadership in the world's aviation community. Now we are working to develop new systems and to enhance a culture that increases the safety, reliability, efficiency, capacity, and environmental performance of our aviation system."*

> Ultimately, UAS must be integrated into the NAS without reducing existing capacity, decreasing safety, negatively impacting current operators, or increasing the risk to airspace users or persons and property on the ground any more than the integration of comparable new and novel technologies.

The FAA created the Unmanned Aircraft Systems Integration Office to facilitate integration of UAS safely and efficiently into the NAS. Toward that goal, the FAA is collaborating with a broad spectrum of stakeholders, which includes manufacturers, commercial vendors, industry trade associations, technical standards organizations, academic institutions, research and development centers, governmental agencies, and other regulators. Ultimately, UAS must be integrated into the NAS without reducing existing capacity, decreasing safety, negatively impacting current operators, or increasing the risk to airspace users or persons and property on the ground any more than the integration of comparable new and novel technologies. Significant progress has been made toward UAS-NAS integration, with many challenges and opportunities ahead.

A key activity of the FAA is to develop regulations, policy, procedures, guidance material, and training requirements to support safe and efficient UAS operations in the NAS, while coordinating with relevant departments and agencies to address related key policy areas of concern such as privacy and national security. Today, UAS are typically given access to airspace through the issuance of Certificates of Waiver or Authorization (COA) to public operators and special airworthiness certificates in the experimental category for civil applicants. Accommodating UAS operations by the use of COAs and special airworthiness certificates will transition to more routine integration processes when new or revised operating rules and procedures are in place and UAS are capable of complying with them. The FAA has a proven certification process in place for aircraft that includes establishing special conditions when new and unique technologies are involved. This process will be used to evaluate items unique to UAS. In those parts of the NAS that have demanding communications, navigation, and surveillance performance requirements, successful demonstration of UAS to meet these requirements will be necessary.

The process of developing regulations, policy, procedures, guidance material, and training requirements, is resource-intensive. This roadmap will illustrate the significant undertaking it is to build the basis for the NAS to transition from UAS *accommodation* to UAS *integration*. Government and industry stakeholders must work collaboratively and apply the necessary resources to bring this transition to fruition while supporting evolving UAS operations in the NAS.

The purpose of this roadmap is to outline, within a broad timeline, the tasks and considerations needed to enable UAS integration into the NAS for the planning purposes of the broader UAS community. The roadmap also aligns proposed Agency actions with the Congressional mandate in the *FAA Modernization and Reform Act of 2012,* Pub. L. 112-95. As this is the first publication of this annual document, the FAA will incorporate lessons learned and related findings in subsequent publications, which will include further refined goals, metrics, and target dates.

The FAA is committed to the safe and efficient integration of UAS into the NAS, thus enabling this emerging technology to safely achieve its full potential.

Purpose and Background of Civil UAS Roadmap

Unmanned aircraft systems (UAS) and operations have significantly increased in number, technical complexity, and sophistication during recent years without having the same history of compliance and oversight as manned aviation. Unlike the manned aircraft industry, the UAS community does not have a set of standardized design specifications for basic UAS design that ensures safe and reliable operation in typical civilian service applications. As a result, the UAS community often finds it difficult to apply existing FAA guidance. In some cases, interpretation of regulations and/or standards may be needed to address characteristics unique to UAS. Ultimately, the pace of integration will be determined by the ability of industry, the user community, and the FAA to overcome technical, regulatory, and operational challenges. The purpose of this roadmap is to outline, within a broad timeline, the tasks and considerations needed to enable UAS integration into the National Airspace System (NAS) for the planning purposes of the broader UAS community. The roadmap also aligns proposed Agency actions with the Congressional mandate in the *FAA Modernization and Reform Act of 2012,* Pub. L. 112-95.

This five-year roadmap, as required by the *FAA Modernization and Reform Act of 2012* (FMRA), is intended to guide aviation stakeholders in understanding operational goals and aviation safety and air traffic challenges when considering future investments. The roadmap is organized into three perspectives that highlight the multiple paths used to achieve the milestones outlined, while focusing on progressive accomplishments. These three perspectives—*Accommodation, Integration,* and *Evolution*—transcend specific timelines and examine the complex relationship of activities necessary to integrate UAS into the NAS. These three perspectives will be explored in more detail in Section 2.2.4.

Although the FMRA requires a five-year UAS roadmap, it is important to view UAS-NAS integration not only in terms of near-term activities and objectives, but also in the context of mid- and long-term timeframes. The timeframes used in this roadmap are defined in the President's National Aeronautics Research and Development Plan, which specifies less than 5 years as the near-term, 5-10 years as the mid-term, and greater than 10 years as the long-term. For this roadmap, the long-term is defined as

> To gain full access to the NAS, UAS need to be able to bridge the gap from existing systems requiring accommodations to future systems that are able to obtain a standard airworthiness certificate

2022-2026, which is consistent with the Joint Planning and Development Office (JPDO) *National Airspace System Concept of Operations and Vision for the Future of Aviation and NextGen Air Transportation System Integrated Plan.*

Integration of UAS into the NAS will require: review of current policies, regulations, environmental impact, privacy considerations, standards, and procedures; identification of gaps in current UAS technologies and regulations, standards, policies, or procedures; development of new technologies and new or revised regulations, standards, policies, and procedures; and the associated development of guidance material, training, and certification of aircraft systems, propulsion systems, and airmen. The FAA will coordinate these integration activities with other United States Government agencies, as needed, through the Interagency Planning Committee (IPC).

1.1 History of UAS

Historically, unmanned aircraft have been known by many names including: "drones," "remotely piloted vehicles (RPV)," "unmanned aerial vehicles (UAV)," "models," and "radio control (R/C) aircraft." Today, the term UAS is used to emphasize the fact that separate system components are required to support airborne operations without a pilot onboard the aircraft. Early UAS operations received little attention from the FAA and its predecessor agencies due to the infrequency of operations, which were mostly conducted in remote locations or in special use airspace and were not deemed to impact the safety of the NAS. In the past two decades, the number of unmanned aircraft operations has been increasing dramatically, highlighting the need for a structured approach for safe and efficient integration.

1.2 Proposed Civil and Commercial Applications

The use of UAS in commercial applications is expected to expand in a number of areas (see Operational Services and Environment Definition (OSED) for Unmanned Aircraft Systems (UAS), RTCA DO-320, 2010). Some of the currently proposed civil and commercial applications of UAS include:

- Security awareness;
- Disaster response, including search and support to rescuers;
- Communications and broadcast, including news/sporting event coverage;
- Cargo transport;
- Spectral and thermal analysis;
- Critical infrastructure monitoring, including power facilities, ports, and pipelines;
- And commercial photography, aerial mapping and charting, and advertising.

1.3 Definitions

Several terms used in this document are defined below as a common point of reference:

Unmanned Aircraft (UA): A device used or intended to be used for flight in the air that has no onboard pilot. This device excludes missiles, weapons, or exploding warheads, but includes all classes of airplanes, helicopters, airships, and powered-lift aircraft without an onboard pilot. UA do not include traditional balloons (see 14 CFR Part 101), rockets, tethered aircraft and un-powered gliders.

Crewmember [UAS]: In addition to the crewmembers identified in 14 CFR Part 1, a UAS flightcrew member includes pilots, sensor/payload operators, and visual observers (VO), but may include other persons as appropriate or required to ensure safe operation of the aircraft.

Unmanned Aircraft System (UAS): An unmanned aircraft and its associated elements related to safe operations, which may include control stations (ground, ship, or air-based), control links, support equipment, payloads, flight termination systems, and launch/recovery equipment. As shown in Figure 1, it consists of three elements:

- Unmanned Aircraft;
- Control Station;
- And Data Link.

National Airspace System (NAS): The common network of U.S. airspace — air navigation facilities, equipment, and services; airports or landing areas; aeronautical charts, information and services; rules, regulations, and procedures; technical information; and manpower and material. (see Figure 2)

Next Generation Air Transportation System (NextGen): According to the FAA's *Destination 2025,* (2011):

> *"NextGen is a series of inter-linked programs, systems, and policies that implement advanced technologies and capabilities to dramatically change the way the current aviation system is operated. NextGen is satellite-based and relies on a network to share information and digital communications so all users of the system are aware of other users' precise locations."*

Unmanned Aircraft System (UAS)

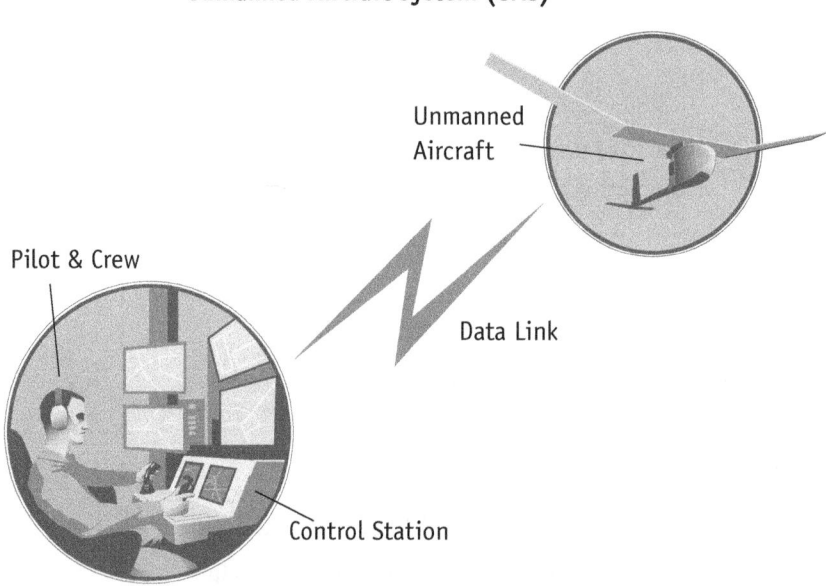

Figure 1: The UAS and Flightcrew Members

1.4 Policy

The FAA is responsible for developing plans and policy for the safe and efficient use of the United States' navigable airspace. This responsibility includes coordinating efforts with national security and privacy policies so that the integration of UAS into the NAS is done in a manner that supports and maintains the United States Government's ability to secure the airspace and addresses privacy concerns. Further, the FAA will harmonize, when appropriate, with the international community for the mutual development of civil aviation in a safe and orderly manner. Components of existing FAA and International Civil Aviation Organization (ICAO) policy are outlined below.

National Airspace System

Figure 2: The NAS

1.4.1 FAA UAS Policy Basis

Established FAA aviation policies support an acceptable level of safety for the NAS. At the core of these policies is the concept that each aircraft is flown by a pilot in accordance with required procedures and practices. This same policy applies to UAS.

Aviation policies and regulations focus on overall safety being addressed through three primary areas: equipment, personnel, and operations and procedures. Each of these areas has standards and minimum levels of safety that must be met, independent of each other. As a matter of regulation, for example, a new civil aircraft must be able to independently obtain an airworthiness certificate, regardless of the airspace class where it might be flown. However, as a result or part of this certification, new procedures may be required for flightcrew members and air traffic control (ATC) in order to maintain the minimum level of safety of the NAS while accommodating the new technology. Under special certifications and authorizations, limited operations may be authorized for equipment unable to meet current standards.

The application of these established aviation policies to UAS is summarized in the following key points excerpted from the FAA Notice of Policy: Unmanned Aircraft Operations in the National Airspace System (72 Fed. Reg. 6689 (Feb. 13, 2007)):

- Regulatory standards need to be developed to enable current technology for unmanned aircraft to comply with Title 14 Code of Federal Regulations;

- In order to ensure safety, the operator is required to establish the UAS airworthiness either from FAA certification, a Department of Defense (DoD) airworthiness statement, or by other approved means;

- Applicants also have to demonstrate that a collision with another aircraft or other airspace user is extremely improbable;

- And the pilot-in-command concept is essential to the safe operation of manned operations. The FAA's UAS guidance applies this pilot-in-command concept to unmanned aircraft and includes minimum qualification and currency requirements.

These policies have enabled the accommodation of UAS into the NAS on a limited basis on the foundation that operations are conducted safely, present an acceptable level of risk to the general public, and do no harm to, or adversely impact, other users. To gain full access to the NAS, UAS need to be able to bridge the gap from existing systems requiring accommodations to future systems that are able to obtain a standard airworthiness certificate. These UAS will also need to be flown by a certified pilot in accordance with existing, revised, or new regulations and required standards, policies, and procedures.

1.4.2 International Civil Aviation Organization (ICAO) Policy

ICAO, a special agency of the United Nations, promotes "the safe and orderly development of international civil aviation throughout the world. It sets standards and regulations necessary for aviation safety, security, efficiency, and regularity, as well as aviation environmental protection."

The goal of ICAO in addressing unmanned aviation is to provide the fundamental international regulatory framework to support routine operation of UAS throughout the world in a safe, harmonized, and seamless manner comparable to that of manned operations. Current ICAO guidance material for UAS is published in ICAO Circular 328, "Unmanned Aircraft Systems (UAS) Circular," which provides basic guidelines for Member States to introduce and integrate UAS into airspace in a consistent manner, to ensure global interoperability and regulatory compatibility, when possible. The document's guiding policy on UAS is:

> "A number of Civil Aviation Authorities (CAA) have adopted the policy that UAS must meet the equivalent levels of safety as manned aircraft... In general, UAS should be operated in accordance with the rule governing the flight of manned aircraft and meet equipment requirements applicable to the class of airspace within which they intend to operate...To safely integrate UAS in non-segregated airspace, the UAS must act and respond as manned aircraft do. Air Traffic, Airspace and Airport standards should not be significantly changed. The UAS must be able to comply with existing provisions to the greatest extent possible."

ICAO develops Standards and Recommended Practices (SARP), which are generally followed by national civil aviation authorities of the Member States. The United States is an ICAO Member State, and the FAA plans to harmonize with international efforts and adhere to ICAO SARPs when possible.

1.4.3 Industry Policy Recommendations

RTCA, Inc. is a private, not-for-profit corporation that develops consensus-based recommendations regarding communications, navigation, surveillance, and air traffic management system issues. RTCA functions as a Federal Advisory Committee, and the FAA considers RTCA recommendations when making policy, program, and regulatory decisions. RTCA Special Committee 203 (SC-203) was established in 2004 to help assure the safe, efficient, and compatible operation of UAS with other aircraft operating within the NAS. This Special Committee has developed and documented guiding principles for UAS integration, which are summarized below:

- UAS must operate safely, efficiently, and compatibly with service providers and other users of the NAS so that overall safety is not degraded;
- UAS will have access to the NAS, provided they have appropriate equipage and the ability to meet the requirements for flying in various classes of airspace;
- Routine UAS operations will not require the creation of new special use airspace, or modification of existing special use airspace;
- Except for some special cases, such as small UAS (sUAS) with very limited operational range, all UAS will require design and airworthiness certification to fly civil operations in the NAS;
- UAS pilots will require certification, though some of the requirements may differ from manned aviation;
- UAS will comply with ATC instructions, clearances, and procedures when receiving air traffic services;
- UAS pilots (the pilot-in-command) will always have responsibility for the unmanned aircraft while it is operating;
- And UAS commercial operations will need to apply the operational control concept as appropriate for the type of operation, but with different functions applicable to UAS operations.

Through an FAA-established UAS Aviation Rulemaking Committee (ARC), the FAA continues to collaborate with government and industry stakeholders for recommendations regarding the path toward integration of UAS into the NAS. This effort will harmonize with the work being done by international organizations working toward a universal goal of safe and efficient UAS airspace operations.

1.4.4 Privacy and Civil Liberties Considerations

The FAA's chief mission is to ensure the safety and efficiency of the entire aviation system. This includes manned and unmanned aircraft operations. While the expanded use of UAS presents great opportunities, it also raises questions as to how to accomplish UAS integration in a manner that is consistent with privacy and civil liberties considerations.

As required by the FMRA, the FAA is implementing a UAS test site program to help the FAA gain a better understanding of operational issues relating to UAS. Although the FAA's mission does not include developing or enforcing policies pertaining to privacy or civil liberties, experience with the UAS test sites will present an opportunity to inform the dialogue in the IPC and other interagency forums concerning the use of UAS technologies and the areas of privacy and civil liberties.

As part of the test site program, the FAA will authorize non-federal public entities to establish and operate six test sites in the United States. The FAA recognizes that there are privacy considerations regarding the use of UAS at the test sites. To ensure that these concerns are taken into consideration at the test sites, the FAA plans to require each test site operator to establish a privacy policy that will apply to operations at the test site. The test site's privacy policy must be publicly available and informed by Fair Information Practice Principles. In addition, each site operator must establish a mechanism through which the operator can receive and consider comments on its privacy policy.

The privacy requirements proposed for the UAS test sites are specifically designed for the operation of the test sites and are not intended to predetermine the long-term policy and regulatory framework under which UAS would operate. However, the FAA anticipates that the privacy policies developed by the test site operators will help inform the dialogue among policymakers, privacy advocates, and the industry regarding broader questions concerning the use of UAS technologies in the NAS.

1.4.5 National Security Issues

Integrating public and civil UAS into the NAS carries certain national security implications, including security vetting for certification and training of UAS-related personnel, addressing cyber and communications vulnerabilities, and maintaining/enhancing air defense and air domain awareness capabilities in an increasingly complex and crowded airspace. In some cases, existing security frameworks applied to manned aircraft may be applicable. Other security concerns may require development of new frameworks altogether. The FAA will continue to work with relevant United States Government departments and agencies, and with stakeholders through coordinating bodies such as the IPC and JPDO, to proactively address these areas of concern.

UAS Operations in the NAS

This roadmap focuses on civil UAS access to the NAS. To this end, the FAA and the UAS community are working to address the myriad challenges associated with this effort.

2.1 FAA's Dual Role for UAS Integration

For UAS, as with all aircraft, the FAA acts in a dual role. As the regulator, the FAA ensures aviation safety of persons and property in the air and on the ground. As the service provider, the FAA is responsible for providing safe and efficient air traffic control services in the NAS and the other portions of global airspace delegated to the United States by ICAO.

As part of its regulator role, the Office of Aviation Safety (AVS) efforts are led by the UAS Integration Office. The main focus of the UAS Integration Office is to provide, within the existing AVS structure, subject matter expertise, research, and recommendations to develop policy, regulations, guidance, and procedures for UAS airworthiness and operations in support of safe integration of UAS into the NAS.

As the service provider, the Air Traffic Organization (ATO) efforts are led by the Air Traffic Emerging Technologies Group, which considers operational authorizations for UAS flights that are unable to meet current regulations and procedures. A Certificate of Waiver or Authorization (COA) is issued with limitations and provisions that mitigate the increased risks resulting from the use of uncertified technology. The ATO is responsible for the safe and efficient handling of aircraft and the development of the airspace rules, procedures, and air traffic controller training to support routine operations in the NAS.

2.2 UAS Challenges

A number of issues that impact the integration of UAS into the NAS are being considered across the regulatory and service provider roles of the FAA. To ensure the FAA meets the goals set forth in this roadmap, these offices will be addressing the challenges as outlined in the following subsections.

2.2.1 Policy, Guidance, and Regulatory Product Challenges

To ensure the FAA has the appropriate UAS framework, many policy, guidance, and regulatory products will need to be reviewed and revised to specifically address UAS integration into the NAS. UAS technology and operations will need to mature, and new products may be required in order to meet applicable regulations and standards. Figure 3 depicts policy, guidance, and regulatory product areas requiring research and development. This information is derived from the RTCA notional architecture and is primarily related to airmen and UAS certification.

2

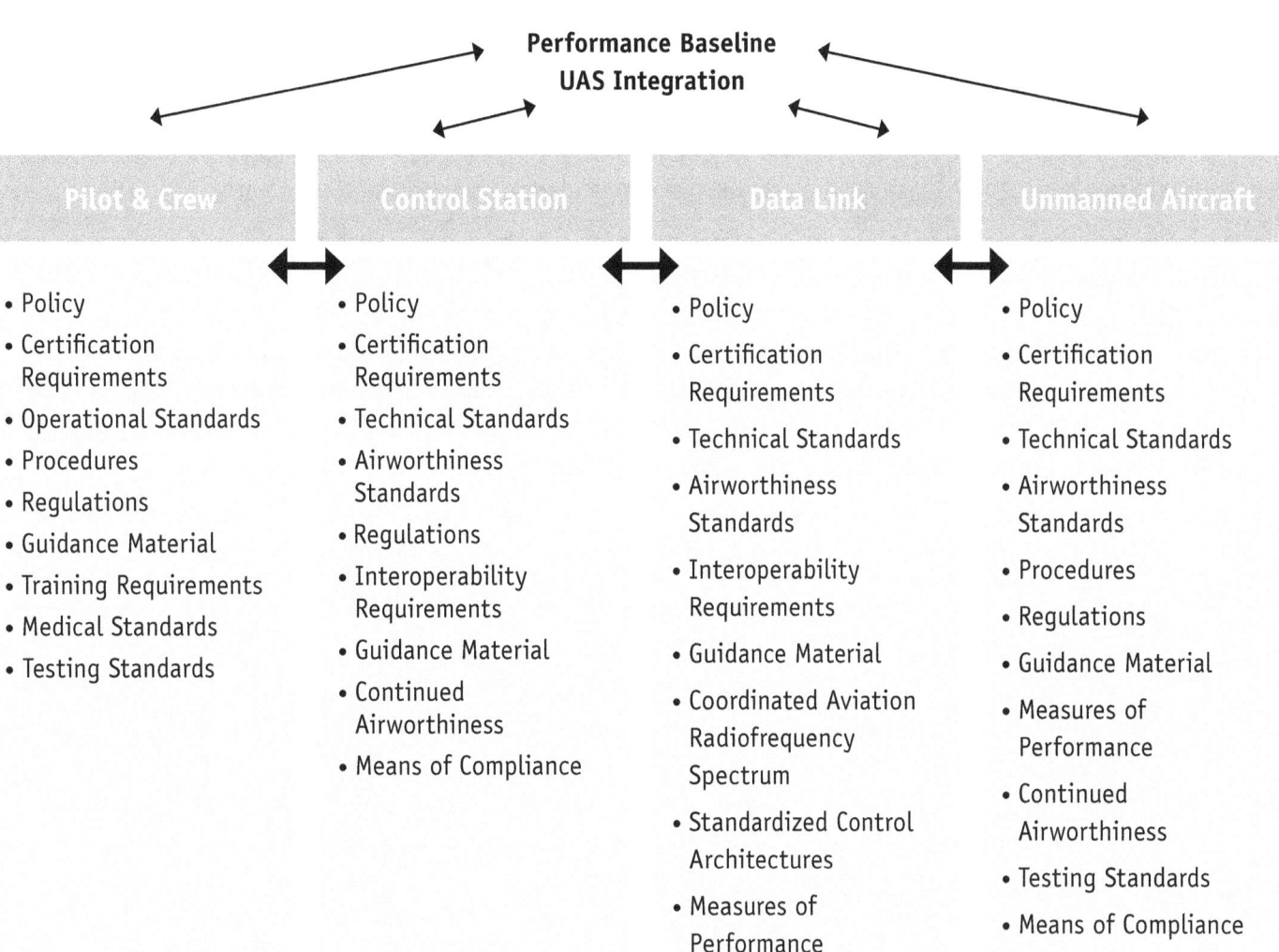

Performance Baseline
UAS Integration

Pilot & Crew	Control Station	Data Link	Unmanned Aircraft

Pilot & Crew
- Policy
- Certification Requirements
- Operational Standards
- Procedures
- Regulations
- Guidance Material
- Training Requirements
- Medical Standards
- Testing Standards

Control Station
- Policy
- Certification Requirements
- Technical Standards
- Airworthiness Standards
- Regulations
- Interoperability Requirements
- Guidance Material
- Continued Airworthiness
- Means of Compliance

Data Link
- Policy
- Certification Requirements
- Technical Standards
- Airworthiness Standards
- Interoperability Requirements
- Guidance Material
- Coordinated Aviation Radiofrequency Spectrum
- Standardized Control Architectures
- Measures of Performance
- Radio/DataLink Security Requirements

Unmanned Aircraft
- Policy
- Certification Requirements
- Technical Standards
- Airworthiness Standards
- Procedures
- Regulations
- Guidance Material
- Measures of Performance
- Continued Airworthiness
- Testing Standards
- Means of Compliance

Figure 3: AVS Products to Regulate UAS Operations

The challenge is to identify and develop the UAS regulatory structure that encompasses areas listed in Figure 3. Other regulatory drivers include:

• Developing minimum standards for Sense and Avoid (SAA), Control and Communications (C2), and separation assurance to meet new or existing operational and regulatory requirements for specified airspace;

• Understanding the privacy, security, and environmental implications of UAS operations and working with relevant departments and agencies to proactively coordinate and align these considerations with the UAS regulatory structure;

• And developing acceptable UAS design standards that consider the aircraft size, performance, mode of control, intended operational environment, and mission criticality.

Although aviation regulations have been developed generically for all aircraft, until recently these efforts were not done with UAS specifically in mind. This presents certain challenges because the underlying assumptions that existed during the previous efforts may not now fully accommodate UAS operations. As an example, current regulations address security requirements for cockpit doors. However, these same regulations lack a legal definition for what a "cockpit" is or where it is located. This presents a challenge for UAS considering that the cockpit or "control station" may be located in an office building, in a vehicle, or outside with no physical boundaries. Applying current cockpit door security regulations to UAS may require new rulemaking, guidance, or a combination of both.

The regulatory process is designed to provide transparency to the public and an opportunity to understand and comment on proposed rules before being issued. Additional checks and balances are in place to ensure that final regulations are not unnecessarily burdensome to the public. Because of these requirements, and lacking any exceptions, an average regulatory effort might span a number of years. These timeframes may be longer for high visibility or complex regulations. FAA experience to date with the development of a Notice of Proposed Rulemaking (NPRM) for small UAS indicates that UAS rulemaking efforts may be more complex, receive greater scrutiny, and require longer development timeframes than the average regulatory effort.

Numerous Air Traffic products, policies, and procedures also need to be reviewed and refined or developed through supporting research to permit UAS operations in the NAS. The UAS Integration Office coordinates efforts with the ATO to complete these tasks.

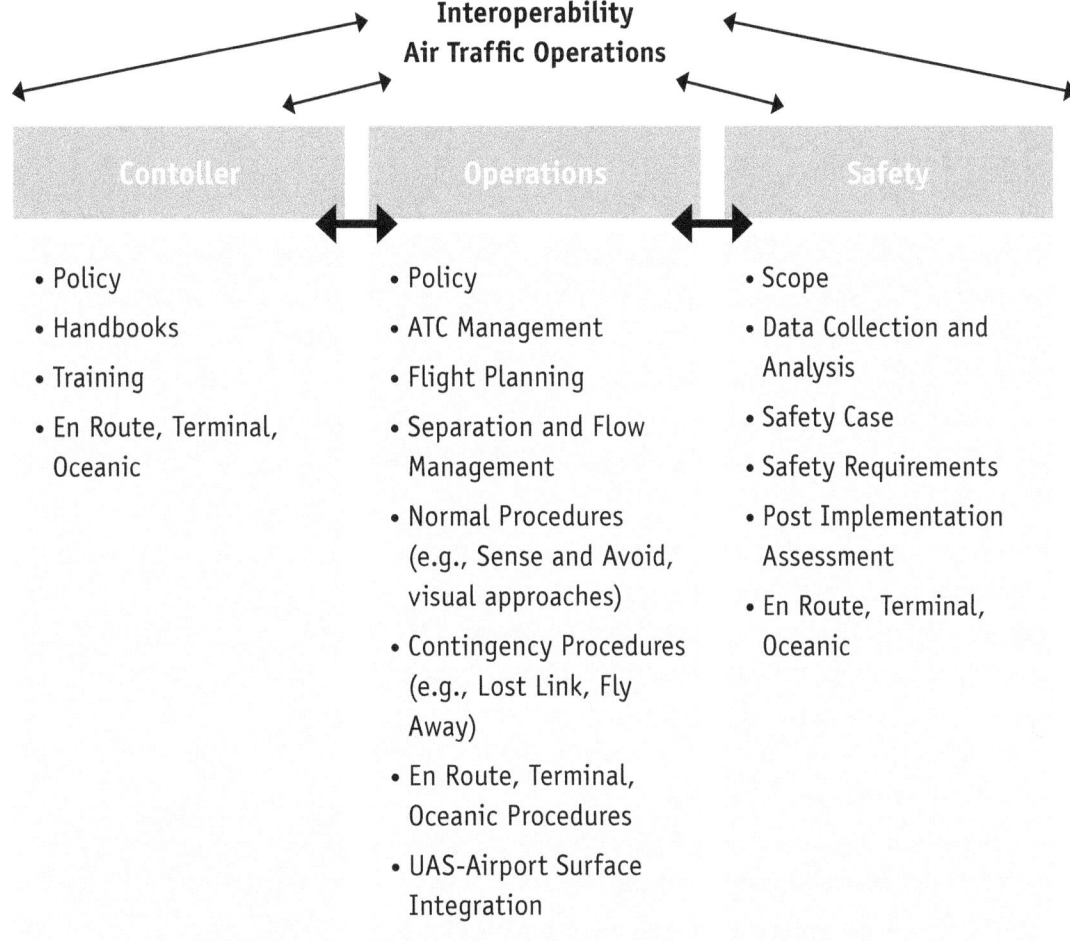

Figure 4: ATO UAS Operational Area

The goal of safely integrating UAS without segregating, delaying, or diverting other aircraft and other users of the system presents significant challenges in the areas outlined in Figure 4 above. For NAS integration, this also includes:

• Identifying policies and requirements for UAS to comply with ATC clearances and instructions commensurate with manned aircraft (specifically addressing the inability of UAS to comply directly with ATC visual clearances or to operate under visual flight rules);

• Establishing procedures and techniques for safe and secure exchange of voice and data communication between UAS pilots, air traffic controllers, and other NAS users;

• Establishing wake vortex and turbulence avoidance criteria needed for UAS with unique characteristics (e.g., size, performance, etc.);

• And reviewing environmental requirements (e.g., the National Environmental Policy Act).

The FAA recognizes that current UAS technologies were not developed to comply with existing airworthiness standards. Current civil airworthiness regulations may not consider many of the unique aspects of UAS operations. Materials properties, structural design standards, system reliability standards, and other minimum performance requirements for basic UAS design need to be evaluated against civil airworthiness standards for existing aircraft. Although significant technological advances have been made by the UAS community, critical research is needed to fully understand the impact of UAS operations in the NAS. There has also been little research to support the equipment design necessary for UAS airworthiness certification. In the near- to mid-term, UAS research will need to focus on technology deemed necessary for UAS access to the NAS.

As UAS are introduced, their expected range of performance will need to be evaluated for impact on the NAS. UAS operate with widely varying performance characteristics that do not necessarily align with manned aircraft performance. They vary in size, speed, and other flight capabilities. Similarly, the issue of performance gap between the pilot and the avionics will impact NAS operations. For example, a quantitative time standard for a pilot response to ATC directions (such as "turn left heading 270, maintain FL250") does not exist – there is an acceptable delay for the pilot's verbal response and physical action, but there is no documented required range of acceptable values. Avionics that perform the corresponding function cannot be designed and built without these performance requirements being established.

Existing standards ensure safe operation by pilots actually on board the aircraft. These standards may not translate well to UAS designs where pilots are remotely located off the aircraft. Removing the pilot from the aircraft creates a series of performance considerations between manned and unmanned aircraft that need to be fully researched and understood to determine acceptability and potential impact on safe operations in the NAS. These include the following considerations:

- The UAS pilot is not onboard the aircraft and does not have the same sensory and environmental cues as a manned aircraft pilot;

- The UAS pilot does not have the ability to directly comply with see-and-avoid responsibilities and UAS SAA systems do not meet current operational rules;

- The UAS pilot must depend on a data link for control of the aircraft. This affects the aircraft's response to revised ATC clearances, other ATC instructions, or unplanned contingencies (e.g., maneuvering aircraft);

- UAS cannot comply with certain air traffic control clearances, and alternate means may need to be considered (e.g., use of visual clearances);

- UAS present air traffic controllers with a different range of platform sizes and operational capabilities (such as size, speed, altitude, wake turbulence criteria, and combinations thereof);

Removing the pilot from the aircraft creates a series of performance considerations between manned and unmanned aircraft that need to be fully researched and understood to determine acceptability and potential impact on safe operations in the NAS.

- And some UAS launch and recovery methods differ from manned aircraft and require manual placement and removal from runways, a lead vehicle for taxi operations, or dedicated launch and recovery systems.

Therefore, it is necessary to develop new or revised regulations/ procedures and operational concepts, formulate standards, and promote technological development that will enable manned and unmanned aircraft to operate cohesively in the same airspace. Specific technology challenges include two critical functional areas:

- **"Sense and Avoid" (SAA) capability** must provide for self-separation and ultimately for collision avoidance protection between UAS and other aircraft analogous to the "see and avoid" operation of manned aircraft that meets an acceptable level of safety. SAA technology development is immature. In manned flight, see and avoid, radar, visual sighting, separation standards, proven technologies and procedures, and well-defined pilot behaviors combine to ensure safe operation. Unmanned flight will require new or revised operational rules to regulate the use of SAA systems as an alternate method to comply with "see and avoid" operational rules currently required of manned aircraft. SAA system standards must be developed to assure both self-separation and collision avoidance capability for UAS. Interoperability constraints must also be defined for safe and secure interactions between SAA-enabled UAS and other airborne and ground-based collision avoidance systems. While SAA may be an independent system, it must be designed to be compatible across other modes (e.g., ATC separation services). See Appendix C.3 and C.4 for specific goals and metrics.

- **Control and Communications (C2) system performance requirements** are needed and RTCA is developing consensus-based recommendations for the FAA to consider in C2 policy, program, and regulatory decisions. The resulting C2 requirements need to support the minimum performance required to achieve higher-level (UAS level) performance and safety requirements. Third-party communication service providers are common today (e.g., ARINC, Harris, etc.) and the FAA has experience with setting and monitoring performance of third parties. The use of third parties is dependent on the UAS architecture chosen, but these are still being evaluated in terms of feasibility from a performance, cost, and safety perspective. See Appendix C.5 for specific goals and metrics.

Unmanned flight will require new or revised operational rules to regulate the use of SAA systems as an alternate method to comply with "see and avoid" operational rules currently required of manned aircraft.

To provide the UAS community insight into the FAA process for fostering UAS flight in the NAS, Figure 5 highlights the intended shift in focus over time from Accommodation to Integration, and then to Evolution. This method is consistent with the approach used for new technologies on manned aircraft introduced into the NAS.

Current design standards reflect the focus in the COA process on allowing existing designs, embodying some experimental design philosophies, to fly in the NAS. Progress toward standard airworthiness will also increase as design standards mature, but not before.

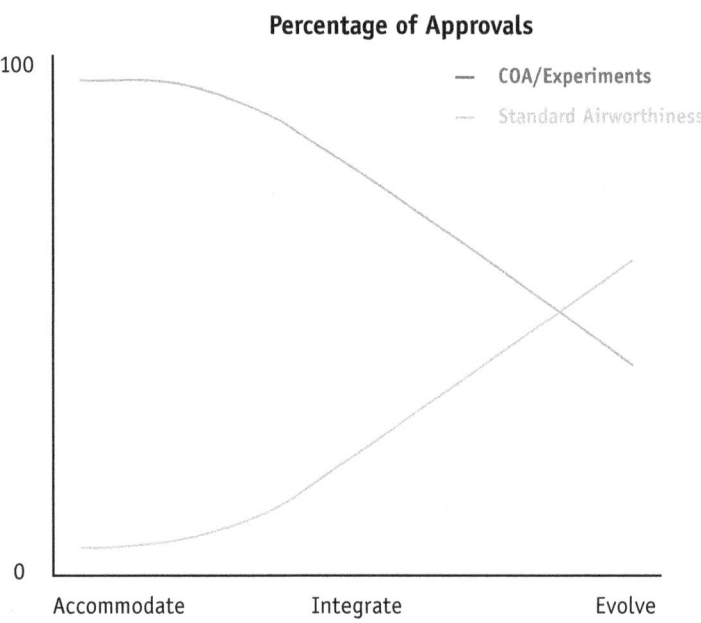

Figure 5: Transition from COA/Experimental to Standard Airworthiness Approvals

Recognizing the challenges and the complex coordination required for integration, the UAS roadmap addresses the efforts needed to move forward incrementally toward the goal of full NAS integration.

Timely progress on products, decisions, research, development, testing, and evaluation will be needed to successfully move from accommodation to integration in the evolving NAS.

The approach to managing the challenges discussed in this section focuses on the following interdependent topics:
• Standards;
• Rules and Regulations;
• Certification of the UAS;
• Procedures and Airspace;
• Training (Pilot, Flightcrew Member, Mechanic, and Controller);
• And Research and Development (R&D) and Technology.

The roadmap discusses the activities and transitions for the above interdependent topic areas from the vantage point of Accommodation, Integration, and Evolution, as summarized below and described in more detail in subsequent sections of this roadmap. These perspectives transcend the near-, mid-, and far-term timeframes and provide additional insight into the task of integrating UAS into the NAS.

Perspective 1: Accommodation. Take current UAS and apply special mitigations and procedures to safely facilitate limited access to the NAS. UAS operations in the NAS are considered on a case-by-case basis. Accommodation will predominate in the near-term, and while it will decline significantly as integration begins and expands in the mid-term, it will continue to be a viable means for NAS access with appropriate restrictions and constraints to mitigate any performance shortfalls. During the near-term, R&D will continue to identify challenges, validate advanced mitigation strategies, and explore opportunities to progress UAS integration into the NAS.

Perspective 2: Integration. Establishing threshold performance requirements for UAS that would increase access to the NAS is a primary objective of integration. During the mid- to far-term, the Agency will establish new or revised regulations, policies, procedures, guidance material, training, and understanding of systems and operations to support routine NAS operations. Integration is targeted to begin in the near- to mid-term with the implementation of the sUAS rule and will expand further over time (mid- and far-term) to consider wider integration of a broader field of UAS.

Perspective 3: Evolution. All required policy, regulations, procedures, guidance material, technologies, and training are in place and routinely updated to support UAS operations in the NAS operational environment as it evolves over time. It is important that the UAS community maintains the understanding that the NAS environment is not static, and that there are many improvements planned for the NAS over the next 13-15 years. To avoid obsolescence, UAS developers will need to maintain a dual focus: integration into today's NAS while maintaining cognizance of how the NAS is evolving.

Perspective 1: Accommodation

3.1 Overview

The FAA's near-term focus will be on safely allowing for the expanded operation of UAS through accommodation. Enhanced procedures and technology, over time, will increase access to the NAS through accommodation made possible by improvements to current mitigations and the introduction of advanced mitigations. The need to maintain this avenue for NAS access will continue. Research and development on current and advanced mitigations is necessary to maintain this avenue for access with appropriate restrictions and constraints to mitigate performance shortfalls and address privacy, security, and environmental concerns. The consideration and planning for integration of UAS into the NAS will continue simultaneously.

There has been a growing interest in a wide variety of civil uses for unmanned aircraft. A number of paths can be used to apply for airworthiness certification of UAS. One method that the UAS civil community is currently using to access the NAS is with a special airworthiness certificate in the experimental category, which requires specific, proven capabilities to enable operations at a constrained level. Each application is reviewed for approval on a case-by-case basis that allows a carefully defined level of access that is limited and dependent on risk mitigations that ensure safety and efficiency of the NAS is not diminished. The use of special airworthiness certificates for UAS is similar to their use for manned aircraft and they are normally issued to UAS applicants for the purposes of research and development, crew training or market surveys per 14 CFR 21.191(a), (c), and (f).

Through August 2012, the FAA had issued 114 special airworthiness certificates (i.e., 113 experimental certificates and one special flight permit) to 22 different models of civil aircraft. Of these 22 different models, 16 are unmanned aircraft and 6 are Optionally Piloted Aircraft (OPA). These experimental certificates have been useful for UAS research and development (R&D), and as R&D efforts subside, the use of experimental certificates may decrease. While the FAA continues to accommodate special access to the NAS, existing airworthiness standards are also an avenue for full-type certification. The FAA is working with the UAS ARC to gain feedback to potential changes to airworthiness standards for UAS, as necessary. In the long-term, UAS that are designed to a standard and built to conform to the design may be integrated into the NAS as fully certificated aircraft.

3

3.2 Standards

If UAS are to operate routinely in the NAS, they must conform to an agreed-upon set of standards. Requirements will vary depending on the nature and complexity of the operation, aircraft or component system limitations, pilot and other crewmember qualifications, and the operating environment.

A technical (or operational) standard is an established norm or requirement about a technical (or operational) system that documents uniform engineering or technical criteria, methods, processes, and practices. A standard may be developed privately or unilaterally, by a corporation, regulatory body, or the military. Standards can also be developed by organizations such as trade unions and associations. These organizations often have more diverse input and usually develop voluntary standards that may be adopted by the FAA as a means of regulatory compliance.

To operate an aircraft safely and efficiently in today's NAS, a means of complying with applicable parts of Title 14 of the Code of Federal Regulations must be developed. Aircraft certification standards govern the design, construction, manufacturing, and continued airworthiness of aircraft used in private and commercial operations. These standards were developed with an underlying assumption that a person would be onboard the aircraft and manipulating the controls. This has led to numerous requirements that make aircraft highly reliable and safe for their intended operations and flightcrew protection.

While UAS share many of the same design considerations as manned aircraft, such as structural integrity and performance, most unmanned aircraft and control stations have not been designed to comply with existing civil airworthiness or operational standards. Beyond the problem of meeting existing aircraft certification standards, other components of the UAS, such as the equipment and software associated with the data link (control and communications) and the launch and recovery mechanisms, are not currently addressed in civil airworthiness or operational standards.

Since 2004, the FAA has developed close working relationships with several standards development organizations. Most of these organizations plan to complete their UAS standards development efforts in the near- to mid-term timeframe. When accepted, these standards development products may provide a means of compliance for rules established in the mid-term. The FAA has also been either the lead or an important participant in cross-agency efforts that influence standards development and has coordinated and harmonized these activities with international efforts such as the ICAO UAS Study Group.

Standardization efforts have already produced a number of useful definitions, guidance documents, and considerations that provide common understanding and add insight and data to UAS integration efforts:

- RTCA/SC-203's Guidance Material (DO-304) and numerous position papers
- RTCA/SC-203's Operational Services and Environment Definition For Unmanned Aircraft Systems (OSED, DO-320), which documents definitions and operating scenarios for different UAS operations in the NAS
- RTCA Air Traffic Management Advisory Committee, Requirements and Planning Work Group Report "Airspace Considerations for UAS Integration in the National Airspace System," March 26, 2008
- SAA Workshop Reports that have documented SAA timelines and definitions

Standards development will continue with the goal of producing Minimum Aviation System Performance Standards (MASPS) by the end of the near-term. RTCA products will be taken under consideration by the FAA in the development of policy and guidance products such as Advisory Circulars. Minimum Operational Performance Standards (MOPS) may be used to define Technical Standard Orders (TSO) in the mid- to long-term timeframe.

Additional coordination and input from the stakeholder community (industry and trade associations, manufacturers, academia, research organizations, and public agencies) is being provided with the recent establishment of the UAS ARC.

Although the need to develop standards cannot be overstated, detailed policy, guidance, technical performance requirements, and operational procedures are also needed to enable manned and unmanned aircraft to fly safely and efficiently in the NAS. See Appendix C for specific goals and metrics.

3.3 Rules and Regulations

Unmanned aircraft operations have significantly increased in number, technical complexity, and sophistication during recent years without specific regulations to address their unique characteristics. For a person wishing to design, manufacture, market, or operate a UAS for a commercial mission and seeking FAA approval for that aircraft, its pilot and the operations, existing rules have not been fully tailored to the unique features of UAS.

The FAA has published a Notice which replaced the previous interim operational guidance material used to support UAS accommodation. Since accommodation is not envisioned to be eliminated entirely, this Notice will need to be updated periodically, even as progress continues simultaneously on development of UAS rules and regulations for integration.

The FAA is also developing an NPRM to allow sUAS to conduct operations. This rulemaking effort includes an associated industry effort to develop consensus standards needed for rule implementation. Assuming the sUAS NPRM effort proceeds to a final rule, associated guidance will also be completed to allow the FAA to approve operations and civil and public UAS operators to apply for and safely implement these sUAS operations. All sUAS rule development and implementation will be in accordance with the FMRA.

During this period, the appropriate regulations are also being reviewed for applicability to UAS operations by the FAA, industry groups, and the

> The emphasis will be on the need for new or revised rules for UAS to operate under instrument flight rules (IFR), including rules to allow UAS operations analogous to manned aircraft using visual capabilities.

UAS ARC. The results of this review will determine any regulatory gaps that need to be addressed in the development of specific UAS guidance and rulemaking. The emphasis will be on the need for new or revised rules for UAS to operate under instrument flight rules (IFR), including rules to allow UAS operations analogous to manned aircraft using visual capabilities. Based on the findings of this review, a determination will be made regarding the need to modify, supplement, or create specific new regulations to support UAS beyond the near-term. UAS rulemaking will follow these steps.

3.4 Airworthiness Certification of the UAS

Airworthiness certification is a process that the FAA uses to ensure that an aircraft design complies with the appropriate safety standards in the applicable airworthiness regulations. FAA type design approval indicates the FAA has evaluated the safety of the unmanned aircraft design and all its systems, which is more rigorous than simply making a determination that the UAS is airworthy.

Airworthiness standards for existing aircraft are codified in Title 14 of the Code of Federal Regulations, with processes described for FAA type certification in FAA Order 8110.4 and airworthiness certification in FAA Order 8130.2. The FAA has the authority and regulations in place to tailor the design standards to specific UAS applications, and plans to use this authority until further experience is obtained in addressing the design issues that are unique to UAS.

Civil UAS are currently accommodated with experimental certificates under FAA Order 8130.34. The FAA and the UAS industry will need to work together to move away from the existing experimental or expendable design philosophy, toward a design philosophy more consistent with reliable and safe civilian operation over populated areas and in areas of manned aircraft operation.

Existing airworthiness standards have been developed from years of operational safety experience with manned aircraft and may be too restrictive for UAS in some areas and inadequate in others. For example, existing structural requirements that ensure safe operation in foreseeable weather conditions that are likely to be encountered represent an example of well-established design requirements that existing UAS designs will most likely need to consider. Structural failures have nearly been eliminated from manned aircraft operations and must be mitigated to a similar level of likelihood in UAS operations.

Detailed consideration of UAS in the certification process will be limited in number until such time as a broad and significant consideration is given to existing standards, regulations, and policy. This will be facilitated by UAS manufacturers making application for type design approval to the FAA. For type design approval, UAS designers must show they meet acceptable safety levels for the basic UAS design, and operators must employ certified systems that enable compliance with standardized air traffic operations and contingency/emergency procedures for UAS.

The FAA believes that the UAS community will be best served by the use of an incremental approach to gaining type-design and airworthiness approval. This incremental approach (see Figure 6) could involve the following steps:

- First, allowing existing UAS designs to operate with strict airworthiness and operational limitations to gain operational experience and determine their reliability in very controlled circumstances, as under the existing COA concept or through regulations specific to sUAS;
- Next, developing design standards tailored to a specific UAS application and proposed operating environment. This step would enable the development of useful unmanned aircraft and system design and operational standards for the UAS to facilitate safe operation, without addressing all potential UAS designs and applications. This would lead to type certificates (TC) and production certificates with appropriate limitations documented in the aircraft flight manual;

• And lastly, defining standards for repeatable and predictable FAA type certification of a UAS designed with the redundancy, reliability, and safety necessary to allow repeated safe access to the NAS, including seamless integration with existing air traffic.

Because the UAS community is well established under its current operational assumptions, it is unlikely the FAA or UAS industry will establish an entire set of design standards from scratch. As additional UAS airworthiness options are considered and UAS airworthiness design and operational standards are developed, type certification may be more efficiently and effectively achieved. The UAS industry will continue to build capabilities into the mid- and long-term timeframes. See Appendix C.1 for specific goals and metrics.

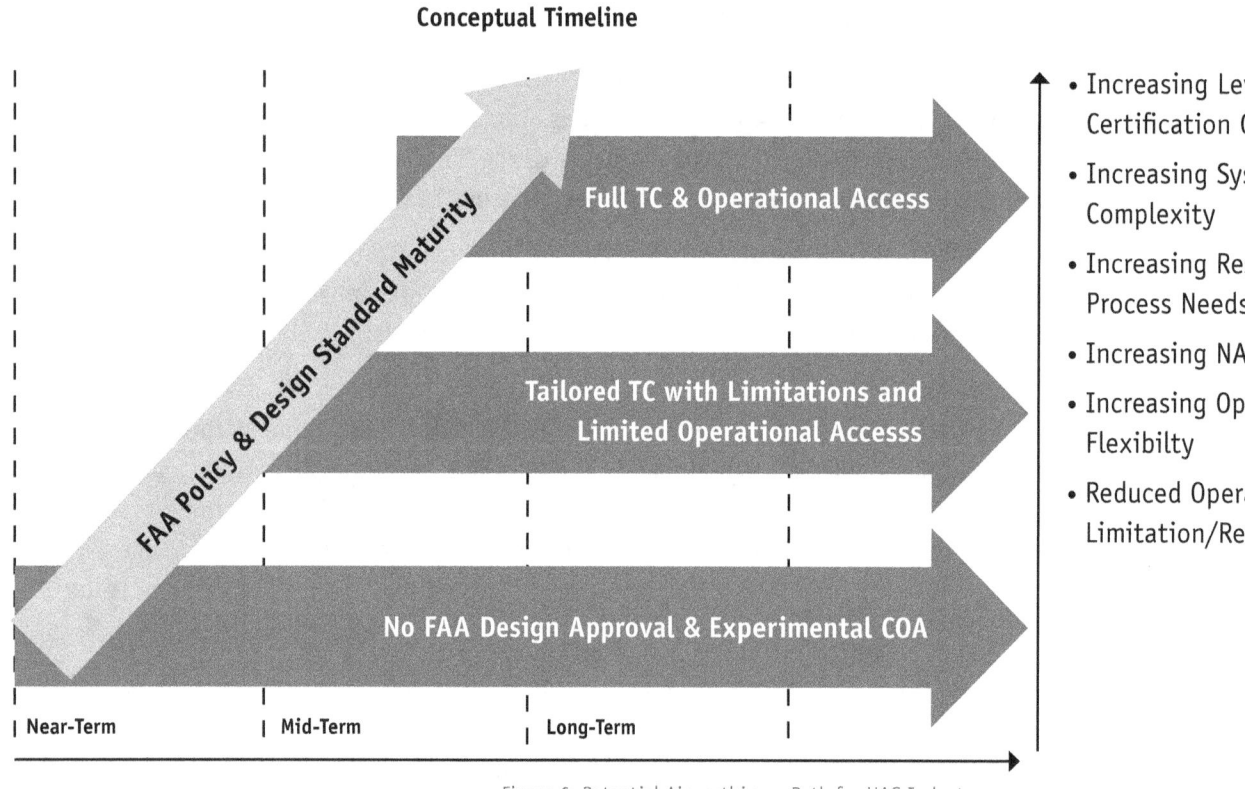

Figure 6: Potential Airworthiness Path for UAS Industry

3.5 Procedures and Airspace

A procedure is a series of actions or operations that have to be executed in the same manner to always obtain the same result under the same circumstances (for example, emergency procedures). The NAS depends on the structure of its airspace and the use of standard procedures to enable safe and efficient operations. ATO directives and other FAA policy and guidance define how UAS are permitted to operate in the NAS today:

• COAs for public access to the NAS – Notice 8900.207 has been released for these operations;

• Experimental Certificates for civil access to the NAS;

• AND AC 91-57 for modeler (recreation) access to the NAS (June 1981) and Section 336 (Special Rule for Model Aircraft) of FMRA.

Experimental certificates and COAs will always be viable methods for accessing the NAS, but typically come with constraints and limitations. Expanded, easier access to the NAS will occur after new or revised operational rules and UAS certification criteria are defined and the FAA develops specific methods for appropriately integrating UAS into NAS operations.

Another requirement is the baselining activity to assess the applicability of existing air traffic control regulations and orders to UAS operations. Any identified gaps will need to be analyzed, and decisions on accommodation or changes to UAS or regulations will be completed. Some sample differences that affect UAS interoperability with the air traffic system are:

• En Route—Current UAS are not able to meet requirements to fly in reduced vertical separation minimum (RVSM) airspace. They do not fly traditional trajectory-based flight paths and require non-traditional handling in emergency situations.

• Terminal—UAS cannot comply with ATC visual separation clearances and cannot execute published instrument approach procedures.

• Facilities—The introduction of UAS at existing airports represents a complex operational challenge. For the near-term, it is expected that UAS will require segregation from mainstream air traffic, possibly accommodated with UAS launch windows, special airports, or off-airport locations where UAS can easily launch and recover. Initial rulemaking for UAS may not address the requirements for UAS at airport facilities, since sUAS are not expected to routinely use airports for takeoff and landing. However, as civil UAS are developed that require airport access, airport integration requirements will need to be developed. These requirements will include environmental impact and/or assessments (when required) concerning noise, emissions, and any unique fuels and other associated concerns. The current Airport Cooperative Research Project (ACRP 03-30) will address the impacts of commercial UAS on airports. The results of the study will be a publication to help airports and communities gain an understanding of UAS, including a description of how various areas of the aviation system, particularly airports, could be affected. The results should be helpful in addressing the airport integration requirement.

ICAO has issued guidance requiring Member States to implement Safety Management System (SMS) programs. These programs are essential to manage risk in the aviation system. The FAA supports this and is a leader in the design and implementation of SMS. Technical challenges abound, including the ability to analyze massive amounts of data to provide useful information for oversight and assessment of risk.

A key input to a Safety Management methodology is the use of safety data. Valuable data collection is underway, but development of a safety-reporting database is currently limited to reporting requirements from existing COAs and experimental certificate holders. Data collection will expand when additional agreements are finalized for sharing public UAS data and new rules and associated safety data reporting requirements are implemented for sUAS. The strategy will use UAS incident, accident, and operational data from public, experimental, and sUAS operations to iteratively support the basis for and define appropriate UAS operating requirements. The availability and quality of this data may directly determine how fast or slow UAS are integrated into the NAS.

3.6 Training (Pilot, Flightcrew Member, Mechanic, and Air Traffic Controller)

UAS training standards will mirror manned aircraft training standards to the maximum extent possible, including appropriate security and vetting requirements, and will account for all roles involved in UAS operation. This may include the pilot, required crew members such as visual observers or launch and recovery specialists, instructors, inspectors, maintenance personnel, and air traffic controllers. See Appendix C.2 and C.8 for specific goals and metrics.

Accident investigation policies, processes, procedures, and training will be developed near-term, and will be provided to Flight Standards District Offices (FSDO) for implementation. Existing manned procedures will be leveraged as much as possible, though differences will need to be highlighted and resolved (e.g., when an unmanned aircraft accident occurs, there may be a need to impound the control station as well as the aircraft).

3.7 Research and Development (R&D) / Technology

Research in the areas of gaps in current technology and new UAS technologies and operations will support and enable the development of airworthiness and operational guidance required to address new and novel aspects of UAS and associated flight operations. The FAA will continue to establish requirements for flight in the NAS so R&D efforts are not duplicative. Additionally, the FAA's research needs are considered within the JPDO NextGen Research Development and Demonstration Roadmap to prevent overlap and provide opportunities for research collaboration.

R&D efforts with industry support the establishment of acceptable performance limits in the NAS and enable the development of performance parameters for today's NAS, while evaluating future concepts, technologies, and procedures for NextGen. The UAS Technical Community Representative Group (TCRG) is sponsoring broad-based UAS research (SAA, C2, and control station studies) aimed at integration with NextGen and validation of concepts. Near-term expected progress is described here:

Sense and Avoid:
Significant research into SAA methods is underway by both government and industry through a variety of approaches and sensor modes. Specifically the FAA is researching:

• Establishment of Sense and Avoid system definitions and performance levels;

• Assessment of Sense and Avoid system multi-sensor use and other technologies;

• And Minimum Sense and Avoid information set required for collision avoidance maneuvering.

Some public agencies and commercial companies are seeking to develop advanced mitigations, such as Ground Based Sense and Avoid (GBSAA) systems, as a strategy for increased access. Concept-of-use demonstrations are underway at several locations to use GBSAA as a mitigation to see-and-avoid requirements for public UAS COA operators in limited operational areas. GBSAA research and the test evaluations will help develop the sensor, link, and algorithm requirements that could allow GBSAA to function as a partial solution set for meeting the SAA requirement and will help build the overall SAA requirements in the long-term. Additionally, as GBSAA technology matures, GBSAA could be used to provide localized UAS NAS integration in addition to being used as an advanced accommodation tool. See Appendix C.3 for specific goals and metrics.

Research is underway on Airborne Sense and Avoid (ABSAA) concepts. Due to complexity, significant progress in ABSAA is not expected until the mid-term. Research goals for the near-term include a flight demonstration of various sensor modes (electro-optic/infrared, radar, Traffic Alert and Collision Avoidance System (TCAS) and Automatic Dependent Surveillance-Broadcast (ADS-B)). Actual fielding of a standardized ABSAA system is a long-term objective. See Appendix C.4 for specific goals and metrics.

Control and Communications:
A primary goal of C2 research is the development of an appropriate C2 link between the unmanned aircraft and the control station to support the required performance of the unmanned aircraft in the NAS and to ensure that the pilot always maintains a threshold level of control of the aircraft. Research will be conducted for UAS control data link communications to determine values for latency, availability, integrity, continuity, and other performance measures.

UAS contingency and emergency scenarios also require research (e.g., how will a UAS in the NAS respond when the command link is lost either through equipment malfunction or malicious jamming, etc.). This research will drive standards that are being established through:

• Development and validation of UAS control link prototype

• Vulnerability analysis of UAS safety critical communications

• Completion of large-scale simulations and flight testing of initial performance requirements

Spectrum and civil radio frequency (RF) identification requires global coordination. The International Telecommunication Union (ITU) through the 2015 World Radiocommunication Conference (WRC-2015) will consider spectrum for UAS beyond-line-of-sight (BLOS) applications. Within the United States, the Federal Communications Commission (FCC) manages and authorizes all non-federal use of the radio frequency spectrum, including state and local government as well as public safety. The National Telecommunications and Information Administration (NTIA) manages and authorizes all federal use of the radio frequency spectrum. UAS spectrum operations within the United States need either the approval of the FCC or NTIA and shall not transmit without being properly authorized. Government agencies and industry need to investigate link security requirements, such as protection against intended and unintended jamming, RF interference, unauthorized link takeover, and spoofing. See Appendix C.5 for specific goals and metrics.

Modeling and Simulation:
The FAA is working with other government agencies and industry to develop a collaborative UAS modeling and simulation environment to explore key challenges to UAS integration. The near-term modeling goals are to:

• Validate current mitigation proposals;

• Establish a baseline of end-to-end UAS performance measures;

• Establish thresholds for safe and efficient introduction of UAS into the NAS;

• And develop NextGen concepts, including 4-dimensional trajectory utilizing UAS technology.

These modeling and simulation efforts will address NAS integration topics for UAS, such as latency in executing ATC clearances, inability to accept ATC visual clearances or comply with visual flight rules, priority and equity of NAS access, lost link, and flyaway scenarios.

Human Factors:

With the pilot controlling the aircraft from beyond the aircraft, several human factors issues emerge related to both the pilot and ATC, and how they will interact to safely operate unmanned aircraft in the NAS. Human factors issues in manned aviation are well known, but there needs to be further analyses regarding integration of UAS into the NAS. In the near-term, data will be collected to permit analysis of how pilots fly UAS, how controllers provide service involving a mix of manned aircraft and UAS, and how pilots and controllers interact with each other, with the goal of developing pilot, ATC, and automation roles and responsibilities concepts. The JPDO, in collaboration with government, academia, and industry researchers, identified several interrelated research challenges:

• Effective human-automation interaction (level; trust; and mode awareness);

• Pilot-centric ground control station design (displays; sensory deficit and remediation; and sterile cockpit);

• Display of traffic/airspace information (separation assurance interface);

• Predictability and contingency management (lost link status; lost ATC communication; and ATC workload);

• Definition of roles and responsibilities (communication flow among crew, ATC, and flight dispatcher);

• System-level issues (NAS-wide human performance requirements);

• And airspace users' and providers' qualification and training (crew/ATC skill set, training, certification, and currency).

Other research in this phase includes activities to support safety case validation and the associated mitigations. This includes case-by-case assessments to determine the likelihood that a system/operation can achieve an acceptable safety level. The research will consider UAS operational and technical risks including:

• Inability to avoid a collision;

• Inability to maintain positive control;

• Inability to meet the operational environment's expected behavior (e.g., self-separate);

• And Inability to safeguard the public.

Summary of "Accommodation" Priorities
Accommodation of UAS in the NAS through evaluation and improvement of safety mitigations
Work with industry and the ARC to review the operational, pilot, and airworthiness regulations
Development of required standards to support technological solutions to identified operational gaps (MOPS)
Safety case validation for UAS operations in NAS—collect/analyze operational and safety data
Robust research, modeling, and simulation for UAS Sense and Avoid, C2, and human factors

Perspective 2: Integration

4.1 Overview

In the mid-term, emphasis will shift significantly from accommodation to integration. For the residual accommodation requirements, it is expected that operational lessons learned and technological advances will lead to more sophisticated mitigations with increased safety margins. Thus, COAs and experimental certificates will remain avenues for accessing the NAS with appropriate restrictions and constraints. Emphasis will shift toward integration of UAS through the implementation of civil standards for unmanned aircraft pilots and new or revised operational rules, together with necessary policy guidance and operational procedures.

Integration efforts will focus on sequentially developing and implementing the UAS system requirements established by the FAA as a result of R&D and test range outputs:

- Finalize the integrated set of FAA rulemaking, policy, operational guidance, procedures, and standards;

- Define continued airworthiness methodologies;

- Complete training and certification standardization;

- Continue the research and technology development and assessment work that underpins the ability of UAS to operate safely and efficiently in the NAS;

- And address the privacy, security, and environmental implications of UAS operations.

To receive civil certification under existing or adapted/expanded regulations, guidance, and standards, research is needed that will assist in defining the certification basis for unique UAS features. While current regulations, guidance, and standards ensure safe operation of aircraft with pilots in the cockpit, these current regulations may not represent the necessary and sufficient basis for the design criteria and operation of UAS.

Integration efforts will provide a foundation for creating and modifying FAA policies and procedures to permit more routine forms of UAS access and bridge the gap to the long-term goal of developing the policy, guidance, and operational procedures required to enable manned and

> Integration efforts will focus on sequentially developing and implementing the UAS system requirements established by the FAA as a result of R&D and test range outputs.

4

unmanned aircraft to fly together in an environment that meets or exceeds today's level of safety and efficiency. As new UAS evolve, more specific training will be developed for UAS pilots, crew members, and certified flight instructors. See Appendix C.2 for specific goals and metrics.

UAS operations comingled at airports with manned aircraft is one of the more significant challenges to NAS integration. The UAS must be able to operate within airport parameters and comply with the existing provisions for aircraft. As with airspace operational requirements, the airport standards are not expected to change with the introduction of UAS, and their operation must be harmonized in the provision of air traffic services.

The following general requirements and assumptions will pertain to all UAS operations that are integrated into the NAS (with the exception of sUAS operating exclusively within visual line-of-sight (LOS) of the flight crew):

1. UAS operators comply with existing, adapted, and/or new operating rules or procedures as a prerequisite for NAS integration.

2. Civil UAS operating in the NAS obtain an appropriate airworthiness certificate while public users retain their responsibility to determine airworthiness.

3. All UAS must file and fly an IFR flight plan.

4. All UAS are equipped with ADS-B (Out) and transponder with altitude-encoding capability. This requirement is independent of the FAA's rule-making for ADS-B (Out).

5. UAS meet performance and equipage requirements for the environment in which they are operating and adhere to the relevant procedures.

6. Each UAS has a flight crew appropriate to fulfill the operators' responsibilities, and includes a pilot-in-command (PIC). Each PIC controls only one UA.*

7. Autonomous operations are not permitted.** The PIC has full control, or override authority to assume control at all times during normal UAS operations.

8. Communications spectrum is available to support UAS operations.

9. No new classes or types of airspace are designated or created specifically for UAS operations.

10. FAA policy, guidelines, and automation support air traffic decision-makers on assigning priority for individual flights (or flight segments) and providing equitable access to airspace and air traffic services.

11. Air traffic separation minima in controlled airspace apply to UA.

12. ATC is responsible for separation services as required by airspace class and type of flight plan for both manned and unmanned aircraft.

13. The UAS PIC complies with all ATC instructions and uses standard phraseology per FAA Order (JO) 7110.65 and the Aeronautical Information Manual (AIM).

14. ATC has no direct link to the UA for flight control purposes.

* This restriction does not preclude the possibility of a formation of UA (with multiple pilots) or a "swarm" (one pilot controlling a group of UA) from transiting the NAS to/from restricted airspace, provided the formation or swarm is operating under a COA.

** Autonomous operations refer to any system design that precludes any person from affecting the normal operations of the aircraft.

4.2 Standards

After MASPS are completed, the emphasis of standards activities will be geared toward the development of MOPS, which will contribute to the basis for regulatory changes and the equipment standards for UAS-specific systems and equipment. The development of MOPS may provide requirements the FAA may invoke as TSO to support airworthiness approval on certificated unmanned aircraft and may lead to the development of improved systems, potentially applicable to all civil aircraft. See Appendix C for specific goals and metrics.

4.3 Rules and Regulations

Recognizing that the UAS community might be better served by specific rules, the FAA is initially proposing to amend its regulations to adopt specific rules for the operation of sUAS in the NAS. These changes will address the classification of sUAS, certification of sUAS pilots, registration of sUAS, approval of sUAS operations, and sUAS operational limits.

Operations of sUAS under new regulations may have operational, airspace, and performance constraints, but will provide experience for pilots and additional data to inform subsequent rulemaking, standards, and training development for safe and efficient integration of other UAS in the NAS.

When the final rule is published and in effect, it will reduce the need for sUAS operators to conduct operations under either a COA or the constraints of an experimental certificate. This will allow operators and the FAA to shift the focus of resources to solutions that will better enable UAS integration. See Appendix C.6 for specific goals and metrics.

4.4 Airworthiness Certification of the UAS

The FAA will work with the UAS community in defining policy and standards that facilitate agreement on an acceptable UAS certification basis for each applicant. This may involve the development of new policy, guidance, rulemaking, special conditions, and methods of compliance. See Section 3.4 for a more detailed discussion and Appendix C.1 for specific goals and metrics.

> As integration continues, new or revised operational rules and associated standards and policies will allow compliant UAS to access additional airspace throughout the NAS.

4.5 Procedures and Airspace

There will be incremental increases in NAS access based on rigorous safety mitigations of current UAS that were previously developed and built without approved industry or governmental standards. As integration begins, there will be approved airspace and procedures for sUAS, which will provide a basis for developing plans for increased NAS access as UAS are certified. As integration continues, new or revised operational rules and procedures, and associated standards and policies, will allow compliant UAS to access additional airspace throughout the NAS. The ATO will use procedures with these UAS similar to those used for manned aircraft, but may also delegate separation responsibility to UAS for some operations. To support this, ATO goals will be:

- Standardize air traffic operations and contingency/emergency procedures for UAS operators to ensure certified aircraft systems are interoperable with air traffic procedures and airspace requirements;

- Develop airport facility integration plans. This will require research and the development of procedures that address critical issues such as low visibility, taxi spacing, light gun signals, and compatibility with NextGen operations;

- Establish UAS operating requirements with associated ATC procedures for airport conditions;

- And coordinate with the Department of Defense (DoD) and all other appropriate departments and agencies on the development of any new parallel procedures and requirements for air domain awareness and defense.

See Appendix C.8 for specific goals and metrics.

4.6 Training (Pilot, Flightcrew Member, Mechanic, and Air Traffic Controller)

The FAA's role in training is to establish policy, guidance, and standards. Airmen training standards are under development and will be synchronized with the regulatory guidance. Civil operators normally develop a training regimen that allows pilots and flight support to meet regulatory standards. For any UAS operation, training regimens analogous to those that exist for manned aircraft will need to be considered, including relevant areas such as written tests, practical examinations, and currency and proficiency requirements.

Standards for airmen will proceed following the sUAS regulation. The FAA will issue UAS airman certificates and support activities to enable UAS operations to include:

- Development of practical test standards (PTS) and UAS airmen knowledge test question banks;

- Development of a UAS handbook for airmen;

- Training of aviation safety inspectors (ASI) at the FSDO level to provide practical test oversight;

- Identification of designated pilot examiners (DPE) to assist the FSDOs;

- Development of a UAS handbook for pilot and instructors;

- Development of PTS and UAS pilot knowledge test question banks;

- Development of UAS mechanic training and certificate process;

- And development of flight crew security requirements by the relevant United States Government agencies.

Pilot endorsements may be developed for specific UAS makes and models to permit commercial operations. Pilot qualifications by make and model will be built into training and will be expanded based on pilot experience.

Training standards development will be more complex for UAS with unique operating parameters and will continue into the long-term as these UAS are certified.

Regardless of the UAS platform, similar types of training regimens are expected, consisting of a written knowledge test, practical test standards, and a flight evaluation. There will be a requirement for currency and proficiency; qualified ASIs will be fielded to regional offices across the country.

With the introduction of UAS into the NAS, additional training requirements specific to different types of UAS characteristics will probably be required for ATC personnel, including UAS performance, behavior, communications, unique flight profiles, ATC standardized procedures, lost link/fly away profiles, operating limitations, and emergency procedures. Controller training will include differences in interoperability between manned and UAS flights, with a focus on specific handling issues of the aircraft. This training must be administered to ATC facilities throughout the NAS. It is expected that controllers will handle UAS the same as manned aircraft; therefore, no special ATC certification would be required. See Appendix C.2 and C.8 for specific goals and metrics.

4.7 Research and Development (R&D)/Technology

Sense and Avoid:

Research on SAA sensor performance, data communication, and algorithms must provide solutions for safe separation for integration of UAS into the NAS. Research to develop separation algorithms will be accomplished with the JPDO R&D plan goals of:

• Flight demonstration of self-separation and collision avoidance algorithms, with multiple sensors and intruders;

• Assessment of the performance of various self-separation concepts as a function of surveillance data configurations, and evaluation of risk-based self-separation algorithms and policy issues;

• Assessment of the performance of various separation assurance concepts, and flight demonstration of separation assurance algorithms, with criteria-based separation;

• And assessment of UAS performance for delegated spacing applications (e.g., defined interval clearances).

Although research will continue, fully certified UA-based collision avoidance solutions may not be feasible until the long-term and are deemed to be a necessary component for full UAS NAS integration. This will include research on safe and efficient terminal airspace and ground operations, followed by ground demonstrations of autonomous airfield navigation and ATC interaction. See Appendix C.4 and C.8 for specific goals and metrics.

Control and Communications:

Advanced research is required in data link management, spectrum analysis, and frequency management. Efforts will focus on completing development of C2 link assurance and mitigation technologies and methods for incorporating them into the development of certification of the UAS. This will include:

• Identification of satellite communication spectrum from the ITU through its WRC;

• Verification and validation of control communication final performance requirements;

• Establishment of UAS control link national/international standards;

• And development and validation of technologies to mitigate vulnerabilities.

Complete characterization of the capacity, performance, and security impacts of UAS on ATC communication systems will be completed. See Appendix C.5 and C.8 for specific goals and metrics.

Human Factors:

Human factors research will continue in the areas of human-machine interface (both control station displays and ATC displays), automation, and migration of control. Human factors data collected in the near-term and mid-term will be analyzed to determine the safest technologies and best procedures for pilots and ATC controllers to interact with each other and with the aircraft; these results will influence technology and operations research. For separation and collision avoidance capability, the contribution of human decision making versus automation must be identified. See Appendix C.8 for specific goals and metrics.

4.8 Test Ranges

Per the FMRA, the FAA will establish six test ranges. The test ranges will take into consideration climate and geographic diversity, the location of ground infrastructure and research needs. See Appendix C.7 for specific goals and metrics.

The test range program will address and account for:

- Manned-unmanned operations,
- Certification standards and air traffic requirements,
- Coordination and leveraging of National Aeronautics and Space Administration (NASA) and DoD resources,
- Civil and public unmanned aircraft systems,
- And coordination with NextGen.

The test ranges will help provide a verification mechanism for safe operations before unmanned aircraft are integrated into the NAS.

The FAA anticipates test range operator privacy practices, as discussed in their privacy policies, will help inform the dialogue among policymakers, privacy advocates, and the industry regarding broader questions concerning the use of UAS technologies. Transparency of privacy policies associated with UAS test range operations will engage all stakeholders in discussions about which privacy issues are raised by UAS operations and how law, public policy, and the industry practices should respond to those issues in the long run.

Summary of "Integration" Priorities
New operational rules and associated standards, policies, and procedures established for small UAS
New operational rules and associated standards, policies, and procedures established for other UAS
C2 link standards defined for integrity, latency, and continuity
FAA acceptance of MASPS to enable development of detailed MOPS
Published FAA policy and operational guidance to define acceptable methods to comply with operational rules in accordance with an acceptable UAS certification basis for each applicant
Published FAA flightcrew training and certification standards

Pererspectivs 3: Evolution

5.1 Overview

Overlaying the integration of UAS is the need to remain aware of the changing characteristics and requirements of the evolving NAS. The long-term focus for UAS operations is the refinement and updating of regulation, policy, and standards. The end-state is to implement streamlined processes for the continued integration of UAS into the NAS.

These efforts will include:

• Policy, operational guidance, and standards for civil aircraft airworthiness and NAS operations and with consideration for privacy and security concerns and frameworks;

• Continued airworthiness methodologies;

• Training and certification standardization;

• And certification of key technologies to enable continued operations of UAS in the NAS.

5.2 Standards

Unique UAS certification requirements will have been determined. MASPS, MOPS, and TSOs will support the regulations and certification of key systems for each UAS. Additionally, all standards will be evaluated and modified, as needed. See Appendix C.1 for specific goals and metrics.

5.3 Rules and Regulations

Lessons learned from previous rulemaking efforts may be applicable to the development of new UAS regulations. The process should become more efficient as UAS experience is gained and data analysis proves safety cases more quickly. UAS rulemaking activities will be more likely to involve revisions to existing rules, as needed, rather than the creation of new rules.

5.4 Airworthiness Certification of the UAS

Certification of UAS will evolve as future technologies evolve and will be consistent with all other aircraft airworthiness and operational approval processes, adding more capability to the UAS through data analyses and trending, which will identify areas for change and improvement in operations, human factors, communication links, and maintenance. See Section 3.4 for a more detailed discussion and Appendix C.1 for specific goals and metrics.

5

5.5 Procedures and Airspace

Certified pilots and UAS will be permitted access into the NAS under seamless operating procedures. The need to accommodate special NAS access will be dramatically reduced, and will be limited to research and development or test operations.

UAS operations will continue to evolve based on NextGen requirements. See Appendix C.8 for specific goals and metrics.

5.6 Training (Pilot, Flightcrew Member, Mechanic, and Air Traffic Controller)

As new UAS evolve, more specific training will be developed for UAS pilots, crew members, and certified flight instructors based on lessons learned and data collection. See Appendix C.2 and C.8 for specific goals and metrics.

5.7 Research and Development (R&D) / Technology

Identified limitations and gaps will be closed via research and development of required technologies that meet standards established by the FAA. Planned activities include:

• Sense and Avoid research that focuses on algorithm development and compatibility with current and future manned aircraft collision avoidance systems such as TCAS II/ACAS X and surveillance systems (e.g., ADS-B), as well as compatibility with ATC separation management procedures and tools;

• Research on UAS system safety and levels of automation for the improvement of UAS into the future;

• Examination of potential concepts for the widespread integration of UAS into the future NextGen environment;

• AND research on new tools and techniques to support avionics and control software development and certification, to ensure their safety and reliability.

Organized studies will continue to investigate the evolution of UAS operations into the NextGen environment. Detailed research on SAA flight operations, using certified sensor systems, could allow aircraft to maintain safe distances from other aircraft during flight conditions that would not be appropriate for visual flight in a manned aircraft. This capability would rely heavily on network-enabled information, precision navigation, and cooperative surveillance, and would require the development and integration of NextGen-representative technologies for traffic, weather, and terrain avoidance. This conceptual model will be enlarged with sensors that expand the ability to maintain separation from other aircraft past the current visual spectrum and flight conditions restrictions. See Appendix C.8 for specific goals and metrics.

Summary of "Evolution" Priorities

Seamless operations of certified UAS and crew members in the evolving NAS

Published FAA TSOs based on system level MOPS

Certified Sense and Avoid algorithms for collision avoidance and self-separation that are interoperable with evolving NextGen ATC systems and manned collision avoidance systems

Conclusions

6.1 Summary

The safe integration of unmanned aircraft into the NAS is a significant challenge. The FAA is dedicated to developing the technical and regulatory standards, policy guidance, and operational procedures on which successful UAS integration depends.

The application of financial and human resources by academia and industry to support critical FAA initiatives will shorten the time required to develop technical and regulatory standards. Together, all stakeholders can overcome the challenge of integrating UAS into the NAS and leverage UAS and associated technologies for the greater benefit of society.

6.2 Outlook

Based on FAA policy and the challenges that need to be addressed, this roadmap has focused on the activities required to achieve integration of UAS into the evolving NAS. Throughout the process, the key messages below reflect the basis for the FAA's consideration of requirements to integrate civil UAS into the NAS:

1) **Government-industry collaboration is paramount to success and must focus on process, quality, and timely results.**
   The FAA expects to gain experience in applying the existing airworthiness regulations during the type certification process with early UAS adopters. We also expect input from industry and the ARC. Taking into account industry and ARC inputs, and future experience with UAS type certification projects, the FAA will review and revise as necessary the existing airworthiness regulations to ensure UAS safety.

2) **The FAA must remain committed to the development of technical and regulatory standards, policy guidance, and operations procedures on which successful UAS integration depends.**
   With this roadmap, the FAA has outlined initiatives that must be accomplished. Because unmanned aircraft are considered aircraft that are flown by pilots, existing regulations and procedures are largely applicable. However, the complete integration of UAS at airports and in the various airspace classes may necessitate the development of new or revised regulations and supplemental procedures. These will be developed and implemented in coordination with relevant agencies to address related security and privacy implications.

3) **Global standards encourage harmonization and yield cost-effective development.**
   The FAA is not bound by international policies and standards. However, harmonizing efforts with the international aviation community will allow for more seamless operations of UAS across national boundaries. Synchronizing

efforts within the aviation community will also permit better use of limited human and fiscal resources, thereby reducing the time required to produce regulatory guidance, policy, and standards.

4) **The FAA is focused on increased access for UAS without impacting the safety or efficiency of the NAS, while managing environmental impacts.**

The FAA has placed a high priority on the development of rules for small UAS that will increase access to the NAS and provide an initial opportunity for commercial operations. In the long-term, the principal objective of the aviation regulatory framework is to achieve and maintain the highest possible uniform level of safety while maintaining or increasing the efficiency and the environmental performance of the NAS. In the case of UAS, this means ensuring the safety of all airspace users as well as the safety of persons and property on the ground.

5) **Progress must be made on the development of technology to enable NAS access.**

Because of many distinct differences between UAS and manned aircraft, there are required technologies that must be matured to enable the safe and seamless integration of UAS in the NAS. Research will be focused in the areas of sense and avoid, control and communications, and human factors.

Appendix A: Acronyms

ABSAA	Airborne Sense and Avoid	FAA	Federal Aviation Administration
ACAS X	Airborne Collision Avoidance System X	FCC	Federal Communications Commission
ADS-B	Automatic Dependent Surveillance–Broadcast	FMRA	FAA Modernization and Reform Act of 2012
AIM	Aeronautical Information Manual	FSDO	Flight Standards District Office
AMA	Academy of Model Aeronautics	GBSAA	Ground Based Sense and Avoid
ARC	Aviation Rulemaking Committee	GSE	Ground Support Equipment
ASI	Aviation Safety Inspector	IFR	Instrument Flight Rules
ASTM	American Society for Testing and Materials	ICAO	International Civil Aviation Organization
ATC	Air Traffic Control	IPC	Interagency Planning Committee
ATO	Air Traffic Organization	ITU	International Telecommunication Union
AVS	Office of Aviation Safety	JPDO	Joint Planning and Development Office
BLOS	Beyond-Line-of-Sight	LOS	Line-of-Sight
C2	Control and Communications	MASPS	Minimum Aviation System Performance Standard
COA	Certificate of Waiver or Authorization	MOPS	Minimum Operational Performance Standard
DAA	Detect and Avoid	NAS	National Airspace System
DHS	Department of Homeland Security	NASA	National Aeronautics and Space Administration
DoD	Department of Defense	NextGen	Next Generation Air Transportation System
DOJ	Department of Justice	NIJ	National Institute of Justice
DPE	Designated Pilot Examiner	NOAA	National Oceanic and Atmospheric Administration

NPRM	Notice of Proposed Rulemaking	TCRG	Technical Community Representative Group
NTIA	National Telecommunications and Information Administration	TSO	Technical Standard Order
		UAS	Unmanned Aircraft System
OPA	Optionally Piloted Aircraft	UAV	Unmanned Aerial Vehicle
OSED	Operational Services and Environmental Definition	VO	Visual Observer
PIC	Pilot-in-Command	WRC	World Radiocommunication Conference
PTS	Practical Test Standards	14 CFR	Title 14 of the Code of Federal Regulations
R/C	Radio Control		
R&D	Research and Development		
RF	Radio Frequency		
RPV	Remotely Piloted Vehicle		
RVSM	Reduced Vertical Separation Minimum		
SAA	Sense and Avoid		
SARP	Standards and Recommended Practices		
SMS	Safety Management System		
S&T	Science and Technology		
sUAS	Small Unmanned Aircraft Systems		
TC	Type Certificate		
TCAS	Traffic Alert and Collision Avoidance System		

Appendix B: Glossary

The following definitions were obtained from several sources, including:

1. Title 14 of the Code of Federal Regulations, Part 1.1
2. FAA Pilot/Controller Glossary (P/CG)
3. RTCA DO-320: Operational Services and Environmental Definition for Unmanned Aircraft Systems
4. Notice 8900.207, "Unmanned Aircraft Systems (UAS) Operational Approval," January 22, 2013
5. FAA Modernization and Reform Act of 2012
6. "Sense and Avoid (SAA) for Unmanned Aircraft Systems (UAS)" – Second Caucus Workshop Report 2013
7. FAA Order 8130.34B – Airworthiness Certification of Unmanned Aircraft Systems and Optionally Piloted Aircraft

Note: Applicable sources are shown at the end of each definition in parentheses (e.g. (1), (2), etc.). Terms without a specific source definition are defined in this Roadmap.

Terminology	Definition
Air Traffic Control	A service operated by appropriate authority to promote the safe, orderly, and expeditious flow of air traffic. (1)
Aircraft	A device that is used or intended to be used for flight in the air. (1)
Airspace	Any portion of the atmosphere sustaining aircraft flight and which has defined boundaries and specified dimensions. Airspace may be classified as to the specific types of flight allowed, rules of operation, and restrictions in accordance with International Civil Aviation Organization standards or State regulation. (3)
Airworthiness Certification	A process that the FAA uses to ensure that an aircraft design complies with the appropriate safety standards in the applicable airworthiness regulations.
Certificate of Waiver or Authorization	An FAA grant of approval for a specific flight operation. The authorization to operate a UAS in the National Airspace System as a public aircraft outside of Restricted, Warning, or Prohibited areas approved for aviation activities. (4)

B

Terminology	Definition
Civil Aircraft	Aircraft other than public aircraft. (4)
Collision Avoidance	The Sense and Avoid system function where the UAS takes appropriate action to prevent an intruder from penetrating the collision volume. Action is expected to be initiated within a relatively short time horizon before closest point of approach. The collision avoidance function engages when all other modes of separation fail. (6)
Communication Link	The voice or data relay of instructions or information between the UAS pilot and the air traffic controller and other NAS users. (3)
Control Station	The equipment used to maintain control, communicate with, guide, or otherwise pilot an unmanned aircraft. (3)
Crewmember [UAS]	In addition to the crewmembers identified in 14 CFR Part 1, a UAS flightcrew member includes pilots, sensor/payload operators, and visual observers, but may include other persons as appropriate or required to ensure safe operation of the aircraft. (4)
Data Link	A ground-to-air communications system which transmits information via digital coded pulses. (3)
Detect and Avoid	Term used instead of Sense and Avoid in the Terms of Reference for RTCA Special Committee 228. This new term has not been defined by RTCA and may be considered to have the same definition as Sense and Avoid when used in this document.
International Civil Aviation Organization	A specialized agency of the United Nations whose objective is to develop the principles and techniques of international air navigation and to foster planning and development of international civil air transport. (2)
Manned Aircraft	Aircraft piloted by a human onboard. (3)
Model Aircraft	An unmanned aircraft that is capable of sustained flight in the atmosphere; flown within visual line-of-sight of the person operating the aircraft and flown for hobby or recreational purposes. (5)

Terminology	Definition
National Airspace System	The common network of U.S. airspace; air navigation facilities, equipment and services, airports or landing areas; aeronautical charts, information and services; rules, regulations and procedures, technical information, and manpower and material. Included are system components shared jointly with the military. (2)
Optionally Piloted Aircraft	An aircraft that is integrated with UAS technology and still retains the capability of being flown by an onboard pilot using conventional control methods. (7)
Pathfinder	An initial UAS airworthiness certification program that will aid the FAA in the establishment of certification requirements.
Pilot-in-Command	Pilot-in-command means the person who: 1) has final authority and responsibility for the operation and safety of the flight; 2) has been designated as pilot-in-command before or during the flight; and 3) holds the appropriate category, class, and type rating, if appropriate, for the conduct of the flight. (1)
Public Aircraft	An aircraft operated by a governmental entity (including federal, state, or local governments, and the U.S. Department of Defense and its military branches) for certain purposes as described in 49 U.S.C. §§ 40102(a)(41) and 40125. Public aircraft status is determined on an operation by operation basis. See 14 CFR Part 1, § 1.1 for a complete definition of a public aircraft. (4)
RTCA	RTCA, Inc. is a private, not-for-profit corporation that develops consensus-based recommendations regarding communications, navigation, surveillance, and air traffic management system issues. RTCA functions as a Federal Advisory Committee. Its recommendations are used by the FAA as the basis for policy, program, and regulatory decisions and by the private sector as the basis for development, investment and other business decisions. (www.rtca.org)
See and Avoid	When weather conditions permit, pilots operating instrument flight rules or visual flight rules are required to observe and maneuver to avoid another aircraft. Right-of-way rules are contained in 14 CFR Part 91. (2)
Self-Separation	Sense and Avoid system function where the UAS maneuvers within a sufficient timeframe to remain well clear of other airborne traffic. (6)
Sense and Avoid	The capability of a UAS to remain well clear from and avoid collisions with other airborne traffic. Sense and Avoid provides the functions of self-separation and collision avoidance to establish an analogous capability to "see and avoid" required by manned aircraft. (6)
Small Unmanned Aircraft	An unmanned aircraft weighing less than 55 pounds. (5)

Terminology	Definition
Special Airworthiness Certificate – Experimental Category (UAS)	Airworthiness certification for experimental UAS and optionally piloted aircraft.
Test Range	A defined geographic area where research and development are conducted in accordance with Sections 332 and 334 of the FMRA. Test ranges are also known as test sites in related documents such as the FAA's Screening Information Request. (5)
Unmanned Aircraft	1) A device used or intended to be used for flight in the air that has no onboard pilot. This devise excludes missiles, weapons, or exploding warheads, but includes all classes of airplanes, helicopters, airships, and powered-lift aircraft without an onboard pilot. UA do not include traditional balloons (see 14 CFR Part 101), rockets, tethered aircraft and un-powered gliders. (4)
	2) An aircraft that is operated without the possibility of direct human intervention from within or on the aircraft. (5)
Unmanned Aircraft System	An unmanned aircraft and its associated elements related to safe operations, which may include control stations (ground, ship, or air-based), control links, support equipment, payloads, flight termination systems, and launch/recovery equipment. (4)
	An unmanned aircraft and associated elements (including communications links and the components that control the unmanned aircraft) that are required for the pilot-in-command to operate safely and efficiently in the national airspace system. (5)
Visual Line-of-Sight	Unaided (corrective lenses and/or sunglasses exempted) visual contact between a pilot-in-command or a visual observer and a UAS sufficient to maintain safe operational control of the aircraft, know its location, and be able to scan the airspace in which it is operating to see and avoid other air traffic or objects aloft or on the ground. (4)

Appendix C: Goals, Metrics, and Target Dates

This appendix contains FAA-developed goals, metrics, and target dates (date ranges) and incorporates many related Unmanned Aircraft Systems (UAS) Aviation Rulemaking Committee (ARC) recommendations. The target dates in this appendix are generally limited to a five-year planning horizon. The FAA will continue its effective dialogue with the UAS ARC as it makes changes to the existing set of goals, metrics, and target dates in yearly updates to this roadmap. These annual updates will track and report progress, as recommended by the Government Accountability Office.

The following material identifies the key goals and related activities to be accomplished in accommodating, integrating, and evolving UAS operations in the National Airspace System (NAS). The goals are, for the most part, intended to be addressed concurrently. For each goal, a set of metrics (i.e., well-defined milestones with target completion dates) is defined. The metrics help establish and maintain common government and industry expectations, and enable objective assessments of the progress made toward the accomplishment of each goal. The goals and metrics reflect the incremental approach to UAS certification and integration described in this roadmap.

The goals and metrics in and of themselves do not constitute a UAS integration roadmap implementation plan; however, they do establish a set of strategic objectives that can guide the definition of activities, schedules, and resource requirements in such a plan. Many of the goals and metrics are not under the FAA's direct control and are dependent upon industry efforts such as participation in civil UAS standards development activities and execution of initial certification (a.k.a. "Pathfinder") programs to aid the establishment of certification requirements. Goals and metrics addressing FMRA requirements are identified and the FMRA Subtitle B (Unmanned Aircraft Systems) is included as a reference in Appendix D.

Target dates for near-term metrics (i.e., those with dates prior to October 2015) are identified by the calendar quarter and year targeted for metric completion (e.g., "3rd Quarter of 2014" means targeted for completion by the end of September 2014). Mid-term metrics may only have a target year or year range specified. In this case, "2016" means the metric's completion target is the end of calendar year 2016. Far-term metrics are outside the five-year horizon of this roadmap and have no target dates. Target dates shown as "from 201x to 201y" indicate related activity is expected throughout this time period. Unless the target dates are required by law (e.g., FMRA), they are exactly that – targets. They are not commitments, either by the FAA, other government organizations, or industry. The target dates consider ongoing and planned government and industry activities and schedules; however, they are not always constrained by these activities and schedules. Some of the target dates are aggressive and will require additional industry or government resources if they are to be met.

Although this roadmap is focused on the integration of civil UAS in the NAS, some of the recommended goals and metrics address public UAS integration activities – primarily those of the Department of Defense (DoD). Public entities may have their own certification processes, but the requirements typically build upon those established by the FAA for civil aviation. The DoD's significant activities to develop public UAS that meet airspace and regulatory requirements can and should be leveraged in the FAA's establishment of civil UAS certification requirements.

C.1 Certification Requirements (Airworthiness)

Note: The term "Operator" is used here as defined by the FAA for passenger/cargo carrying and other "for hire and compensation" operations. Not all UAS operations conducted for hire and compensation will require an Operator Certificate. One outcome of this effort will be to establish which UAS operations will or will not require an Operator Certificate.

Goal 1: FAA initial certification process established for one or more civil applicants by 2014.

    A. One or more Pathfinder certification projects were defined through government-industry plans (e.g., Project Specific Certification Plans (PSCP)) in the 2nd Quarter of 2013.

- Explanation. Three UAS manufacturers have already applied for type certification and two of these applications were released from delayed sequencing to proceed with restricted category airworthiness certification. Restricted category type certifications for these two applicants have now been completed. Completion of these type certification projects under appropriate, existing certification regulations, will act as a catalyst to establish the process to be used for similar UAS type certification projects. Note: Some UAS type certifications may be in the restricted category with operating restrictions to maintain an equivalent level of safety for the public, but the goal is to certify the respective UAS to meet all integration requirements, if practical.

    B. One or more Pathfinder standard airworthiness certification projects complete initial certification planning by 2014.

- Explanation. If the FAA and one or more industry partners complete initial certification planning as recommended in The FAA and Industry Guide to Product Certification, the groundwork will be in place for an efficient certification project that will help establish the process for similar UAS certification projects. One manufacturer has made application and the project will proceed per FAA sequencing processes.

Goal 2: FAA's initial issue papers for one or more standard airworthiness certification projects are available by 2014.

   A. One or more Pathfinder certification projects underway by the 4th Quarter of 2013.

   • Explanation. One manufacturer has made application for a standard airworthiness certificate and the project will proceed per FAA sequencing processes.

   B. FAA's initial certification issues defined for the certification basis or new and novel systems (e.g., UAS control station, airframe, control system, propulsion system, ground support equipment (GSE), etc.) by the 4th Quarter of 2013.

   • Explanation. The certification basis and any unique requirements for new and novel systems must be established. Requirements can be identified and refined as a result of Pathfinder efforts or publication by standards organizations (e.g., RTCA, Inc., ASTM International).

Goal 3: FAA's unique certification requirements identified through issue papers that have matured for one or more standard airworthiness certification projects by 2015.

   A. FAA's unique certification requirements for new and novel systems (e.g., UAS control station, airframe, control system, propulsion system, GSE, etc.) published by 2015.

   • Explanation. Lessons learned from certification of Pathfinder systems, publication of consensus standards, and additional operational experience gained as a result of small UAS (sUAS) rule publication will provide additional requirement information for future applicants.

   B. One or more Pathfinder standard airworthiness certification projects completed by 2017 if all associated activities are completed per the nominal certification process.

   • Explanation. It is expected that type certifications will be granted only when all requirements have been met under existing rules and requirements and this target date is a best-case scenario.

   C. Other certification programs completed by 2017–2020, based on timely applications and system commonality/complexity.

   • Explanation. Lessons learned from certification activities of Pathfinder systems, publication of consensus standards, and operations under the sUAS rule will provide data and experience to support other certification efforts.

Goal 4: FAA certification requirements updated and systems certified as necessary.

   A. Certification requirements updated as necessary.

   B. UAS certified as necessary.

C.2 Certification Requirements (Pilot/Crew)

Goal 1: FAA certification requirements for pilots and crew members for sUAS classes (including medical requirements, training standards, etc.) published as part of a sUAS rule by 2014 in accordance with the FMRA. Note: These requirements include coordination with other government agencies on security/vetting requirements.

Goal 2: Necessary changes to record keeping systems established as part of a sUAS rule and in accordance with the FMRA.

- Explanation. Once the final requirements are established, some changes to existing record keeping systems will be necessary.

Goal 3: FAA certification requirements for pilots and crew members for UAS classes other than those addressed under the sUAS rule (including medical requirements, training standards, etc.) published by 2014–2017.

C.3 Ground Based Sense and Avoid (GBSAA)

Goal 1: FAA draft Advisory Circular on GBSAA systems and requirements released by 2015.

A. FAA approvals for use of GBSAA at one or more DoD GBSAA test sites granted by the 3rd Quarter of 2015, subject to timely application and completion of Certificate of Waiver or Authorization (COA) or other approval processes.

- Explanation. Use U.S. Army and U.S. Air Force developed solutions at DoD UAS test sites. (Note: These are existing DoD GBSAA test sites, not the new test ranges discussed in Section 4.8 and Appendix C.7.)

B. FAA approvals for use of GBSAA for educational and other public applications granted by 2016–2018, subject to timely application and completion of COA or other approval processes.

- Explanation. As above, but expanded beyond the DoD to include public use at other locations equipped with suitable GBSAA systems.

Goal 2: GBSAA operations fully approved by the FAA for routine use by all aviation, including both public and civil entities (if needed).

A. FAA approvals for use of GBSAA for limited civil applications granted.

- Explanation. As with FAA operational approvals for use of GBSAA at all DoD GBSAA test sites and operational approvals for use of GBSAA for educational and other public applications, expanded approvals are expected to be granted for limited civil use at select locations. These approvals will incorporate relevant data from UAS test site operations with GBSAA.

B. FAA's initial GBSAA certification standards for civil operations established.

- Explanation. Assimilate prior deployment experience for DoD, public, and limited civil use, and develop Minimum Aviation System Performance Standards (MASPS) for GBSAA. These approvals will incorporate relevant data from UAS test site operations with GBSAA.

C. FAA approvals for use of GBSAA for civil applications granted.

D. FAA's final GBSAA certification standards for civil operations established.

E. GBSAA certification standards updated as necessary.

Goal 1: Initial FAA certification of ABSAA that facilitates UAS operations without the requirement for a visual observer by 2016–2020.

A. Initial industry proposal for Sense and Avoid (SAA) implementation, integration, and operation in a Pathfinder program provided by the 2nd Quarter of 2014. (See Appendix C.1 for the Pathfinder program goals and metrics.)

  • Explanation. This industry proposal will address: a) general UAS operations requirements, b) UAS sense-and-avoid requirements for all proposed operations, including proposed launch and recovery sites, c) proposed UAS ABSAA equipage, and d) planned installation and integration of the proposed ABSAA system(s). ("System" includes both hardware and software.)

B. FAA Stage 2 issue paper on UAS SAA implementation in one or more Pathfinder programs completed by the 2nd Quarter of 2015, subject to applicant provision of sufficient information in certification application and ongoing processes.

  • Explanation. An FAA Stage 2 issue paper will provide the "FAA Position" indicating the FAA's concerns, opinions, and actions the applicant is required to accomplish to resolve the issue. This position gives the applicant direction that will enable compliance to the requirements without dictating design

Goal 2: Installation and certification of ABSAA developed to meet industry standards for use by the DoD and other public and civil entities that provide the SAA functions required in the NAS for Classes A, E, and G airspace, and operations approved without the requirement for a visual observer or a COA. Note: the RTCA Program Management Committee established a new Special Committee 228 and working group for Detect and Avoid (DAA). SAA and DAA may be used interchangeably until SC-228 provides a unique definition for DAA. Special Committee 228's Terms of Reference acknowledge that the requirements for UAS DAA in some airspace will require rulemaking.

A. RTCA Operational and Functional Requirements and Safety Objectives (OFRSO) for UAS, Volume 1 was released in the 2nd Quarter of 2013.

  • Explanation. The OFRSO "provides recommendations for UAS system level operational and functional requirements and safety objectives for UAS flown in the United States National Airspace System (NAS) under the rules and guidelines for civil aviation." This document provides a framework to support the development of future UAS performance standards and "will prove useful to designers, manufacturers, installers, service providers and users in the development of future standards."

B. RTCA preliminary Phase 1 Detect and Avoid (DAA) Minimum Operational Performance Standards (MOPS) developed to establish performance standards that can be verified and validated for UAS DAA equipment in specified airspace by the 3rd Quarter of 2015.

  • Explanation. Emphasis for this initial phase will be standards development on civil UAS equipped to operate into Class A airspace under IFR. A second phase of MOPS development may include DAA equipment to support extended UAS operations in Class D, E and perhaps G airspace. This work effort includes recommendations for a verification and validation test program to be completed before the release of the DAA MOPS. Note: RTCA has sunset Special Committee 203 and Special Committee 228; has a new Detect and Avoid working group developing these DAA MOPS.

C. RTCA Phase 1 DAA MOPS released by the 3rd Quarter of 2016.

- Explanation. This document includes the avionics onboard the UAS and required elements of ground control systems and is based on the results of verification and validation activities on the preliminary Phase 1 DAA MOPS.

D. FAA DAA Technical Standard Order (TSO) issued by the 1st Quarter of 2017.

- Explanation. This document includes the avionics onboard the UAS and required elements of ground control systems.

E. FAA DAA TSO-required equipment used operationally.

Goal 3: DoD or other public entity certification of initial ABSAA systems that enable the DoD and other public entities to safely operate ABSAA-equipped UAS in all NAS airspace classes without the need for a COA. Note: RTCA Special Committee 228's Terms of Reference acknowledge that the requirements for UAS DAA in some airspace will require rulemaking.

A. Initial proposal for ABSAA implementation, integration, and operation in one or more programs released by 2016.

- Explanation. This proposal will address the requirements for ABSAA system(s), including the SAA avionics onboard the unmanned aircraft and required elements of ground control systems. "System" includes both hardware and software.

B. FAA issue paper(s) on UAS SAA implementation in one or more programs for UAS operations in one or more airspace classes released.

- Explanation. The FAA issue paper(s) will document the special considerations for certification of UAS airborne systems that include SAA functions. They also will document special considerations for operating UAS that employ these ABSAA systems and special considerations (including avionics equipage requirements) for manned aircraft operating in the same airspace.

Goal 4: Installation and certification of ABSAA systems for use by the DoD and other public and civil entities that provide the SAA functions that facilitate integrated operation of manned and unmanned aircraft in all NAS airspace classes.

A. RTCA OFRSO for UAS, Volume 1 was released in the 2nd Quarter of 2013.

- Explanation. The OFRSO "provides recommendations for UAS system level operational and functional requirements and safety objectives for UAS flown in the NAS under the rules and guidelines for civil aviation." This document provides a framework to support the development of future UAS performance standards.

B. RTCA Phase 1 DAA MOPS released by the 3rd Quarter of 2016.

- Explanation. This document includes the SAA avionics onboard the aircraft and required elements of ground control systems for IFR flight in Class D, E, and G airspace as noted in the Terms of Reference.

C. RTCA DAA MOPS released for other classes of airspace.

- Explanation. The second phase of DAA MOPS may specify DAA equipment to support extended UAS operations in Class D, E, G, and other airspace as noted in the Terms of Reference.

D. FAA initial DAA TSO released by the 1st Quarter of 2017.

- Explanation. This document will include the avionics onboard the aircraft and required elements of ground control systems as invoked from requirements specified in the Phase 1 DAA MOPS.

E. FAA DAA TSO-required equipment used operationally.

- Explanation. UAS will receive operational approval to use DAA equipment through standard operational approval processes that may include exemptions to Part 91 and/or rulemaking activities as defined by FMRA.

F. RTCA UAS OFRSO and DAA MOPS updated as necessary.

G. FFAA DAA TSO(s) updated as necessary.

C.5 Control and Communications (C2)

Note: For purposes of this section, line-of-sight (LOS) means radio LOS, not visual LOS.

Goal 1: International agreements, industry standards, and FAA regulations and guidance material established by 2015 for civil UAS Control and Communications (C2) capabilities such that C2 subsystems can be certified by the FAA for use in FAA-approved UAS operations.

Note: C2 includes communications internal to the UAS for pilots to operate unmanned aircraft from ground control stations.

A. International agreement was reached in February 2012 at the International Telecommunication Union's (ITU) World Radiocommunication Conference (WRC) on spectrum identified for radio LOS UAS C2 links (or in ITU terminology, Control and Non-Payload Communications links).

- Explanation. Internationally harmonized radio spectrum is needed to help ensure protection from unintentional radio frequency interference, to help ensure adequate spectral bandwidth is available, and to facilitate operation of UAS across international borders. While spectrum is also needed for beyond-line-of-sight (BLOS) C2 links, the initial focus was on radio line-of-sight for civil UAS because demand for LOS links is expected to be greater.

B. RTCA OFRSO for UAS, Volume 1 was released in the 2nd Quarter of 2013.

- Explanation. The OFRSO "provides recommendations for UAS system level operational and functional requirements and safety objectives for UAS flown in the NAS under the rules and guidelines for civil aviation." This document provides a framework to support the development of future UAS performance standards.

C. RTCA's initial MOPS for all the UAS subsystems involved in providing or enabling C2 Data Link using L-Band and C-Band Terrestrial data links are available to be verified and validated by the 3rd Quarter of 2015.

- Explanation. These preliminary MOPS and associated recommendations for a verification and validation test program are needed for the FAA and industry to mature the final Terrestrial data link standards before the release of the final MOPS. RTCA is expected to define MOPS that include L-Band and C-Band frequencies identified at WRC 2012. The resulting MOPS form the basis upon which the FAA can certify systems and services used in providing C2 capabilities for civil UAS.

D. RTCA final Phase 1 C2 Terrestrial Data Link MOPS released by the 3rd Quarter of 2016.

- Explanation. These performance standards in both L-Band and C-Band spectrum are based on the results of the verification and validation test program activities. RTCA is expected to define MOPS that include L-Band and C-Band frequencies identified at WRC 2012.

E. FAA's initial regulations and guidance material (such as TSOs and Advisory Circulars) to enable the production, sale, installation, and maintenance of FAA-certified systems and services used in providing radio LOS C2 capabilities for civil UAS published by 2016–2017.

- Explanation. For the commercial marketplace to offer FAA-certified systems and services for use in providing C2 capabilities for civil UAS, the FAA must establish the necessary regulations and guidance material. These are expected to be based on and largely incorporate the consensus industry standards defined in the RTCA MOPS.

F. Initial FAA-certified Terrestrial C2 Data Link subsystems intended for civil UAS operations are available commercially.

- Explanation. FAA-certified Terrestrial C2 Data Link subsystems for civil UAS are needed for operators and manufacturers to incorporate in their UAS, and for operators to obtain FAA approval for their UAS operations.

Goal 2: Beyond-Line-of-Sight C2 links and capabilities are addressed in international agreements, industry standards, and FAA regulations and guidance material.

A. International agreement reached at the ITU's WRC 15 on radio spectrum identified for BLOS UAS C2 links by 2015.

- Explanation. Internationally harmonized radio spectrum is needed for UAS C2 links to help ensure their protection from unintentional radio frequency interference, to help ensure adequate spectral bandwidth is available for meeting the projected C2 link capacity demands, and to facilitate operation of UAS across international borders. In the far-term, an increasing number of civil UAS operations are expected to require BLOS C2 links.

B. RTCA's second phase MOPS for all the UAS subsystems involved in providing or enabling radio BLOS C2 capabilities for civil UAS published. These elements will include the necessary portion of avionics onboard the unmanned aircraft, the voice and data links, and the necessary portion of ground control systems.

- Explanation. This second phase of MOPS will be needed to provide standards for the use of SATCOM in multiple bands as a C2 Data Link to support UAS. This development will be based on the results of the ITU's WRC 15 as well as lessons learned from industry application of the initial MOPS during product development and FAA certification activities.

C. FAA's final regulations and guidance material to enable the production, installation, and maintenance of FAA-certified systems and services used in providing radio BLOS C2 capabilities for civil UAS published.

- Explanation. A revised set of FAA regulations and guidance material will be needed to address BLOS C2 Data Link systems. These regulations and guidance material will apply lessons learned from application of the initial set.

D. Initial FAA-certified BLOS C2 subsystems intended for civil UAS operations are available commercially.

- Explanation. FAA-certified BLOS C2 subsystems for civil UAS are needed for operators and manufacturers to incorporate in their UAS, and for operators to obtain FAA approval for their UAS operations.

Goal 3: Adequate spectrum is available for both radio LOS and BLOS C2 links to meet the current and projected demand generated by civil UAS operations in the NAS.

A. International spectrum identified for LOS and BLOS UAS C2 links reviewed for possible modification at a future WRC by 2020.

C.6 Small UAS (sUAS) and Other Rules

Goal 1: sUAS rule adopted to allow for both civil and public operations.

A. Agreements (Memorandums of Understanding (MOU), Memorandums of Agreement (MOA), COA, etc.) among the FAA and the DoD, the Department of Homeland Security (DHS), the National Aeronautics and Space Administration (NASA), the National Oceanic and Atmospheric Administration (NOAA), the Department of Justice (DOJ) and other public entities finalized and signed in conjunction with the release of the sUAS Notice of Proposed Rulemaking (NPRM). (The sUAS NPRM is expected to be released in early 2014).

- Explanation. The sUAS proposed rule has undergone a risk assessment by the FAA through its Safety Management System (SMS) process. Adopting or applying the provisions of the proposed rule for public operations is necessary and will accelerate NAS integration of sUAS. It will also reduce the number of COAs the FAA will need to process and free up FAA resources to address other time-critical UAS in the NAS integration issues.

B. sUAS follow-on night operations experiments and study accepted by the FAA for review by the 3rd Quarter of 2014.

- Explanation. NASA completed an initial study at New Mexico State University in 2012. The FAA reviewed the report on this initial study and provided questions and other inputs for inclusion in NASA's planned follow-on study. The FAA will review the report of these focused experiments and may consider DoD and other night operational data.

C. If night operations are deemed as safe as or safer by the FAA, increased night operations for public entities are allowed by the 3rd Quarter of 2015.

- Explanation. Public entities are requesting night operations as a means to fully exploit the capability of sUAS.

D. D. Drafts of all required consensus standards necessary for the implementation of 14 CFR Part 107 available to the public in conjunction with the release of the sUAS NPRM (currently expected to be released in 2013).

- Explanation. More than three years of consensus standard development have occurred. When completed, these standards will provide meaningful guidance to manufacturers and end users for the design, construction, and operation of sUAS. The timely release of the standards will permit industry an opportunity to fully prepare for publication of a final rule, and provide useful guidance to public entities desiring UAS deployment prior to final rule release.

Goal 2: sUAS rule adoption for public and civil operations.

A. 14 CFR Part 107 published, consensus-based standards accepted by the FAA, and the FAA able to issue permits to operate in accordance with requirements of the FMRA.

- Explanation. In order for operations to be conducted under 14 CFR Part 107, the FAA will issue a Notice of Applicability of referenced consensus-based standards and will be able to issue permits to operate.

B. Update sUAS rules, guidance, and/or consensus-based standards after sufficient data have been gathered and analyzed.

- Explanation. Assuming a final rule implementation, the FAA will gain experience with sUAS operating under 14 CFR Part 107. Advancements in technology and analysis of operational and safety data will provide the catalyst for refinement and improvement of Part 107 guidance and/or standards.

C. Update sUAS rules, guidance, and/or consensus-based standards as necessary.

- Explanation. As more operational and safety data is accumulated it will provide a catalyst for refinement and improvement of 14 CFR Part 107 guidance and/or standards as necessary.

Goal 3: sUAS rule supports ATC interoperability to ensure safe and efficient NAS operations.

A. Train air traffic control workforce within six months after sUAS rule enactment.

B. Ensure consistency between sUAS rule proposed operational expectations and proposed changes to ATC Handbook and the Aeronautical Information Manual (AIM).

C. sUAS operations are aligned with ATC Handbook and AIM when the sUAS rule is published and effective.

D. Employ existing strategies to conduct UAS integration safety analysis within SMS Manual guidance to ongoing safety analyses supporting ATC interoperability.

- Explanation. The FAA will enhance ATC interoperability under sUAS rule operations with safety analyses, as required.

Goal 4: Other Rulemaking per the FMRA.

A. Notice of Proposed Rulemaking published to implement the recommendations of the plan required by FMRA by the 3rd Quarter of 2014.

- Explanation. Section 332, subsection (a)(1) of the FMRA specifies plan requirements and subsection (b) requires publication of an NPRM.

B. Final rule published to implement the recommendations of the plan required by the FMRA by the 4th Quarter of 2015.

- Explanation. Section 332, subsection (a)(1) of the FMRA specifies plan requirements and subsection (b) requires publication of a final rule not later than 16 months after publication of the associated NPRM.

C. C. Update to the Administration's most recent policy statement on unmanned aircraft systems contained in Docket No. FAA–2006–25714 required by the FMRA by the 3rd Quarter of 2014.

- Explanation. Section 332, subsection (b) requires publication of this update.

Goal 1: FAA program to integrate UAS into the NAS at six test ranges established in accordance with the FMRA.

- Explanation. To establish this program, selection criteria and procedures were developed and communicated to prospective site operators. Test areas criteria consider geographic and climate diversity, the location of ground infrastructure, and research needs. FAA dialogue with prospective site operators clarified criteria and procedures by gathering prospective site operator questions and documenting answers for use by all.

Goal 2: Test ranges selected by FAA in accordance with the FMRA.

- Explanation. The FAA received applications from prospective site operators in the 1st Quarter of 2013 and is evaluating the applications per the established selection criteria and procedures. Any test range selected should provide the FAA, NASA, DoD, industry and academia with the opportunity for UAS prototype development and deployment.

Goal 3: Selected test ranges operational in accordance with the FMRA

- Explanation. The FMRA states that "the test range shall be operational no later than 180 days after the date on which a project is established."

Goal 4: Test range program operational until February 2017.

- Explanation. The FMRA requires the test range program to be terminated by February 2017.

Goal 5: Report findings and conclusions concerning projects in accordance with the FMRA.

- Explanation. The FMRA states that "Not later than 90 days after the date of the termination of the program...the Administrator shall submit to the Committee on Commerce, Science, and Transportation of the Senate and the Committee on Transportation and Infrastructure and the Committee on Science, Space, and Technology of the House of Representatives a report setting forth the Administrator's findings and conclusions concerning the projects."

Goal 1: Safety and Interoperability—The overall level of safety in the NAS is preserved through NAS integration, which requires adherence to rigorous airworthiness standards and airspace regulations. While they apply equally to manned aircraft, they also recognize the distinguishing characteristics of UAS.

- A. Conduct research that validates the required functional and performance capabilities for safe operation of UAS within the various airspaces of the NAS from 2012 to 2017.

- B. Air Traffic interoperability requirements will be allocated to appropriate Air Traffic program and UAS integration efforts from 2012 to 2017.

- C. Employ existing strategies to conduct UAS integration safety analysis within SMS Manual guidance to ongoing safety analyses supporting ATC interoperability.

- D. Conduct research on Sense and Avoid algorithms for collision avoidance and self-separation that are interoperable with evolving Next Generation Air Transportation System (NextGen) ATC systems and manned collision avoidance systems.

E. Analyze human factors data to determine the safest technologies and best procedures for air traffic controllers to provide services to UAS pilots.

F. Track safety and operational data to use as a basis for policy decisions from 2012 to 2017.

Goal 2: Procedures and Training

A. Develop ATC training requirements specific to different types of UAS characteristics, including UAS performance, behavior, communications, unique flight profiles, ATC standardized procedures, lost link/ fly away profiles, operating limitations, and emergency procedures. Initial training produced in 2009, first revision to be available in 3rd Quarter of 2013. Subsequent training development will occur through 2020.

B. B. Administer UAS training to ATC facilities throughout the NAS from 2013 to 2020.

C.9. Miscellaneous

Goal 1: Develop more detailed plans for safely integrating UAS operations in the NAS by 2015.

A. UAS ARC reviewed FAA and industry plans, including the 2006 Airspace Integration Plan, in 2012.

- Explanation. The 2006 Airspace Integration Plan modified the airspace integration plan developed under the government-industry Access 5 program to more directly address the eight major challenges with UAS integration in the NAS. The UAS ARC will review the 2006 plan and update recommendations consistent with current thinking, including goals and metrics documented in this roadmap.

B. UAS ARC made recommendations for changes to FAA and industry programs and provided them to the FAA in the 2nd Quarter of 2013.

- Explanation. The UAS ARC completed its review of FAA and related industry plans and sent the FAA recommendations for additional planning elements and details. These recommendations include proposed changes to existing and planned programs.

C. Updated FAA UAS Integration Roadmap published annually in accordance with the FMRA.

Goal 2: Identify air traffic management system changes required to be implemented in NextGen.

A. UAS are addressed in the FAA's 2012 *NextGen Implementation Plan* by the 4th Quarter of 2013.

- Explanation. This requires explicitly addressing the operation of UAS in the NAS and the evolution of enabling system capabilities in the various NextGen Segment Implementation Plans (NSIP). Although no significant changes to the current NAS and future NextGen are expected for the integration of UAS operations in unrestricted airspace, some system and procedure changes may be necessary. Any changes need to be incorporated in the *NextGen Implementation Plan,* so that appropriate adjustments to program baselines can be made.

B. UAS are addressed in FAA's NextGen Enterprise Architecture by the 4th Quarter of 2013.

- Explanation. This requires explicitly addressing the integration of UAS operations in the NAS, including the necessary operational concepts and system capabilities. The NextGen Enterprise Architecture identifies whatever is needed to integrate UAS operations in unrestricted airspace. The CY 2012 update to the NextGen Enterprise Architecture depicted FMRA milestones in the aircraft roadmap component.

Goal 3: Review and revise and/or develop new UAS operational scenarios to mature UAS operational concept elements, update operational requirements, and validate key concept elements for UAS integration into the NAS.

  A. FAA initiates an effort to review existing UAS operational scenarios/concept elements and revise them and/or develop new scenarios, if needed, for use in UAS operational concept development per established air traffic system engineering practices by the 1st Quarter of 2014.

  - Explanation. A rich set of operational scenarios and mature concept elements is needed to develop a complete set of operational requirements, from which system functional and performance requirements can be derived. Off-nominal operations may also be defined for conceivable contingency situations.

  B. FAA uses vetted operational scenarios and other concept maturation products to update UAS operator and NAS operational requirements by the 3rd Quarter of 2014.

  - Explanation. This process uses vetted scenarios and other mature concept elements to update and document the set of UAS operator and NAS operational requirements associated with integrating UAS operations into the NAS per established air traffic system engineering analyses and related processes.

  C. FAA uses vetted operational scenarios, updated UAS operator and NAS requirements and other mature concept elements to validate key concept elements and requirements associated with integrating UAS operations into the NAS.

  - Explanation. Air traffic system engineering processes continue to validate concept elements and requirements based on priority need for their validation. Concept element validation priorities will determine resource allocations and schedule for validation of respective concept elements.

Goal 4: Develop UAS integration in the Arctic Region in accordance with the FMRA

  A. A. FAA evaluates key operational concepts for potential inclusion into appropriate operational policy and procedures documents (e.g., FAA Order 8900.1 (Flight Standards Information Management System), state Aeronautical Information Publication (AIP) supplements, Notices to Airmen (NOTAM), etc.) by the 2nd Quarter of 2015.

  B. FAA and Arctic UAS operators examine the costs (e.g., aircraft certification, mandatory equipage requirements, etc.) and benefits (i.e., value of main business cases for use) by the 3rd Quarter of 2015.

  C. FAA begins Air Traffic Organization (ATO) process to establish UAS Arctic Areas, including airspace designation and DoD notice, by the 3rd Quarter of 2015.

  D. FAA completes safety studies in accordance with Section 335 of the FMRA by the 3rd Quarter of 2015. (Note: The first safety risk management (SRM) panel for initial projects convened in 2013 and has drafted the associated SRM document.)

  E. FAA develops UAS restricted category special airworthiness certificate standards by the 3rd Quarter of 2015.

  F. FAA reviews planning and approval documents (e.g., COA template, FAA Destination 2025, FAA Flight Plan 2012, other FAA/International Civil Aviation Organization (ICAO) documents) and evaluates or adapts their use for Arctic Area operations by 2015.

G. Begin international UAS scientific experiments (Marginal Ice Zone Observations and Processes Experiment (MIZOPEX)) with NASA, NOAA, and the University of Alaska), commercial UAS photography missions, or other expanded use/demonstration of UAS in accordance with the Arctic Plan by the 3rd Quarter of 2013.

Goal 5: Develop implementation of Common Strategy for DOJ and associated law enforcement, fire, and first responder agency use of sUAS in the NAS in accordance with the FMRA.

Note: Progress on original metrics is documented below along with metrics to be completed.

A. FAA began collaboration with the DHS Science and Technology (S&T) Directorate during the 4th Quarter of 2012 to support FAA testing and evaluation program of sUAS for law enforcement and first responders, with high-level suitability criteria.

B. FAA formally accepted and signed the MOU with the DOJ National Institute of Justice (NIJ) in the 1st Quarter of 2013.

C. FAA established a working group to examine validity of legacy pilot-in-command (PIC) and observer medical qualifications currently stipulated in COA guidelines in the 2nd Quarter of 2013.

D. FAA established a liaison with DOJ NIJ and U.S. Fire Administration on the development of common strategies for the deployment of sUAS technologies in support of fire enforcement agencies in the 2nd Quarter of 2013.

E. FAA established a working group to examine validity and alternatives to PIC certification requirements established in the 2nd Quarter of 2013.

F. FAA established a collaborative working group with DOJ NIJ and federal law enforcement agencies to examine, plan, and develop a nationwide COA process/approval for the Federal Bureau of Investigation, Bureau of Alcohol, Tobacco, Firearms and Explosives, the National Park Service, and other federal law enforcement and emergency management agencies with country-wide jurisdictions in the 2nd Quarter of 2013.

G. FAA incorporates key operational concepts of strategy into a revised law enforcement/first responder-specific COA template by the 4th Quarter of 2013.

H. FAA establishes a collaborative working group with DOJ NIJ and appropriate law enforcement agencies and trade associations to examine, plan, and develop a COA approval process for law enforcement and first responder mutual aid operations by the 4th Quarter of 2013.

I. FAA to establish working group with DOJ NIJ and the DHS S&T on the development of a technical bulletin on the Common Strategy for distribution to law enforcement/first responders across the nation by the 4th Quarter of 2013.

J. FAA commences collaboration and coordination with DOJ NIJ and DHS S&T to support the co-hosting of a DOJ/FAA/DHS-focused sUAS conference to be convened in the 4th Quarter of 2013.

K. FAA establishes a collaborative working group with DOJ NIJ, DHS and appropriate law enforcement associations to examine, plan, and develop guidelines for any law enforcement agency contemplating the use of unmanned aircraft by the 4th Quarter of 2013.

L. FAA establishes working group with DOJ NIJ on the development of a process for the collection of Unmanned Aircraft Aviation Operations Report data from law enforcement agencies by the 4th Quarter of 2013.

M. FAA completes the development of law enforcement and first responder sUAS competency evaluation procedures, safety risk analysis plan (SRAP), and evaluation checklist completed by the 4th Quarter of 2013.

N. FAA assists three different-sized law enforcement agencies in first implementation of the Common Strategy – target date coordinated with the agencies and confident timeline for the agencies – 4th Quarter of 2013.

- Explanation: FAA assistance is planned for one small agency (i.e., less than 100 sworn officers), one medium agency (i.e., 100 to 300 sworn officers), and one large law enforcement agency (i.e., greater than 300 sworn officers).

O. FAA establishes working group to examine sUAS aircraft recommended guidelines for law enforcement agencies contemplating the use of unmanned aircraft by the 4th Quarter of 2013.

P. FAA will complete COA online modifications to enable Common Strategy implementation by law enforcement agencies by the 4th Quarter of 2014.

Q. FAA reviews planning and approval documents (e.g., unique law enforcement agency COA template, FAA Flight Plan 2012, other FAA/ICAO documents) and modifies or adapts their use for law enforcement agency and first responder sUAS operations by 2015.

Goal 6: In accordance with the FMRA, develop policies to ensure "the Administrator of the FAA may not promulgate any rule or regulation regarding a model aircraft, or an aircraft being developed as a model aircraft."

A. Publish FAA order to establish criteria the agency will use to determine which model aircraft organizations can be considered community-based organizations.

B. Publish update to Federal Register that compares content of AC 91-57 and the FMRA, provides examples of careless and reckless operations, and makes distinction between modeling and commercial operations.

Goal 7: Requirements for the operation of "public unmanned aircraft systems" in the NAS in accordance with the FMRA.

A. Develop and implement operational and certification requirements for the operation of "public unmanned aircraft systems" in the NAS by the 4th Quarter of 2015.

Appendix D: FAA Modernization and Reform Act of 2012 Reference Text

Subtitle B—Unmanned Aircraft Systems

SEC. 331. DEFINITIONS.

In this subtitle, the following definitions apply:

(1) ARCTIC.—The term "Arctic" means the United States zone of the Chukchi Sea, Beaufort Sea, and Bering Sea north of the Aleutian chain.

(2) CERTIFICATE OF WAIVER; CERTIFICATE OF AUTHORIZATION.—The terms "certificate of waiver" and "certificate of authorization" mean a Federal Aviation Administration grant of approval for a specific flight operation.

(3) PERMANENT AREAS.—The term "permanent areas" means areas on land or water that provide for launch, recovery, and operation of small unmanned aircraft.

(4) PUBLIC UNMANNED AIRCRAFT SYSTEM.—The term "public unmanned aircraft system" means an unmanned aircraft system that meets the qualifications and conditions required for operation of a public aircraft (as defined in section 40102 of title 49, United States Code).

(5) SENSE AND AVOID CAPABILITY.—The term "sense and avoid capability" means the capability of an unmanned aircraft to remain a safe distance from and to avoid collisions with other airborne aircraft.

(6) SMALL UNMANNED AIRCRAFT.—The term "small unmanned aircraft" means an unmanned aircraft weighing less than 55 pounds.

(7) TEST RANGE.—The term "test range" means a defined geographic area where research and development are conducted.

(8) UNMANNED AIRCRAFT.—The term "unmanned aircraft" means an aircraft that is operated without the possibility of direct human intervention from within or on the aircraft.

(9) UNMANNED AIRCRAFT SYSTEM.—The term "unmanned aircraft system" means an unmanned aircraft and associated elements (including communication links and the components that control the unmanned aircraft) that are required for the pilot in command to operate safely and efficiently in the national airspace system.

SEC. 332. INTEGRATION OF CIVIL UNMANNED AIRCRAFT SYSTEMS INTO NATIONAL AIRSPACE SYSTEM.
  (a) REQUIRED PLANNING FOR INTEGRATION.—

   (1) COMPREHENSIVE PLAN.—Not later than 270 days after the date of enactment of this Act, the Secretary of Transportation, in consultation with representatives of the aviation industry, Federal agencies that employ unmanned aircraft systems technology in the national airspace system, and the unmanned aircraft systems industry, shall develop a comprehensive plan to safely accelerate the integration of civil unmanned aircraft systems into the national airspace system.

   (2) CONTENTS OF PLAN.—The plan required under paragraph (1) shall contain, at a minimum, recommendations or projections on—
      (A) the rulemaking to be conducted under subsection (b), with specific recommendations on how the rulemaking will—
         (i) define the acceptable standards for operation and certification of civil unmanned aircraft systems;
         (ii) ensure that any civil unmanned aircraft system includes a sense and avoid capability; and
         (iii) establish standards and requirements for the operator and pilot of a civil unmanned aircraft system, including standards and requirements for registration and licensing;
      (B) the best methods to enhance the technologies and subsystems necessary to achieve the safe and routine operation of civil unmanned aircraft systems in the national airspace system;
      (C) a phased-in approach to the integration of civil unmanned aircraft systems into the national airspace system;
      (D) a timeline for the phased-in approach described under subparagraph (C);
      (E) creation of a safe
      (F) airspace designation for cooperative manned and unmanned flight operations in the national airspace system;
      (G) establishment of a process to develop certification, flight standards, and air traffic requirements for civil unmanned aircraft systems at test ranges where such systems are subject to testing;
      (H) the best methods to ensure the safe operation of civil unmanned aircraft systems and public unmanned aircraft systems simultaneously in the national airspace system; and
      (I) incorporation of the plan into the annual NextGen Implementation Plan document (or any successor document) of the Federal Aviation Administration.

(3) DEADLINE.—The plan required under paragraph (1) shall provide for the safe integration of civil unmanned aircraft systems into the national airspace system as soon as practicable, but not later than September 30, 2015.

(4) REPORT TO CONGRESS.—Not later than 1 year after the date of enactment of this Act, the Secretary shall submit to Congress a copy of the plan required under paragraph (1).

(5) ROADMAP.—Not later than 1 year after the date of enactment of this Act, the Secretary shall approve and make available in print and on the Administration's Internet Web site a 5-year roadmap for the introduction of civil unmanned aircraft systems into the national airspace system, as coordinated by the Unmanned Aircraft Program Office of the Administration. The Secretary shall update the roadmap annually.

(b) RULEMAKING.—Not later than 18 months after the date on which the plan required under subsection (a)(1) is submitted to Congress under subsection (a)(4), the Secretary shall publish in the Federal Register—

(1) a final rule on small unmanned aircraft systems that will allow for civil operation of such systems in the national airspace system, to the extent the systems do not meet the requirements for expedited operational authorization under section 333 of this Act;

(2) a notice of proposed rulemaking to implement the recommendations of the plan required under subsection (a)(1), with the final rule to be published not later than 16 months after the date of publication of the notice; and

(3) an update to the Administration's most recent policy statement on unmanned aircraft systems, contained in Docket No. FAA–2006–25714.

(c) PILOT PROJECTS.—

(1) ESTABLISHMENT.—Not later than 180 days after the date of enactment of this Act, the Administrator shall establish a program to integrate unmanned aircraft systems into the national airspace system at 6 test ranges. The program shall terminate 5 years after the date of enactment of this Act.

(2) PROGRAM REQUIREMENTS.—In establishing the program under paragraph (1), the Administrator shall—
  (A) safely designate airspace for integrated manned and unmanned flight operations in the national airspace system;
  (B) develop certification standards and air traffic requirements for unmanned flight operations at test ranges;
  (C) coordinate with and leverage the resources of the National Aeronautics and Space Administration and the Department of Defense;
  (D) address both civil and public unmanned aircraft systems;
  (E) ensure that the program is coordinated with the Next Generation Air Transportation System; and (F) provide for verification of the safety of unmanned aircraft systems and related navigation procedures before integration into the national airspace system.

(3) TEST RANGE LOCATIONS.—In determining the location of the 6 test ranges of the program under paragraph (1), the Administrator shall—
  (A) take into consideration geographic and climatic diversity;
  (B) take into consideration the location of ground infrastructure and research needs; and
  (C) consult with the National Aeronautics and Space Administration and the Department of Defense.

(4) TEST RANGE OPERATION.—A project at a test range shall be operational not later than 180 days after the date on which the project is established.

(5) REPORT TO CONGRESS.—

(A) IN GENERAL.—Not later than 90 days after the date of the termination of the program under paragraph (1), the Administrator shall submit to the Committee on Commerce, Science, and Transportation of the Senate and the Committee on Transportation and Infrastructure and the Committee on Science, Space, and Technology of the House of Representatives a report setting forth the Administrator's findings and conclusions concerning the projects.

(B) ADDITIONAL CONTENTS.—The report under sub-paragraph (A) shall include a description and assessment of the progress being made in establishing special use air-space to fill the immediate need of the Department of Defense—

(i) to develop detection techniques for small unmanned aircraft systems; and

(ii) to validate the sense and avoid capability and operation of unmanned aircraft systems.

(d) EXPANDING USE OF UNMANNED AIRCRAFT SYSTEMS IN ARCTIC.—

(1) IN GENERAL.—Not later than 180 days after the date of enactment of this Act, the Secretary shall develop a plan and initiate a process to work with relevant Federal agencies and national and international communities to designate permanent areas in the Arctic where small unmanned aircraft may operate 24 hours per day for research and commercial purposes. The plan for operations in these permanent areas shall include the development of processes to facilitate the safe operation of unmanned aircraft beyond line of sight. Such areas shall enable over-water flights from the surface to at least 2,000 feet in altitude, with ingress and egress routes from selected coastal launch sites.

(2) AGREEMENTS.—To implement the plan under paragraph (1), the Secretary may enter into an agreement with relevant national and international communities.

(3) AIRCRAFT APPROVAL.—Not later than 1 year after the entry into force of an agreement necessary to effectuate the purposes of this subsection, the Secretary shall work with relevant national and international communities to establish and implement a process, or may apply an applicable process already established, for approving the use of unmanned aircraft in the designated permanent areas in the Arctic without regard to whether an unmanned aircraft is used as a public aircraft, a civil aircraft, or a model aircraft.

SEC. 333. SPECIAL RULES FOR CERTAIN UNMANNED AIRCRAFT SYSTEMS.

(a) IN GENERAL.—Notwithstanding any other requirement of this subtitle, and not later than 180 days after the date of enactment of this Act, the Secretary of Transportation shall determine if certain unmanned aircraft systems may operate safely in the national airspace system before completion of the plan and rulemaking required by section 332 of this Act or the guidance required by section 334 of this Act.

(b) ASSESSMENT OF UNMANNED AIRCRAFT SYSTEMS.—In making the determination under subsection (a), the Secretary shall determine, at a minimum—

(1) which types of unmanned aircraft systems, if any, as a result of their size, weight, speed, operational capability, proximity to airports and populated areas, and operation within visual line of sight do not create a hazard to users of the national airspace system or the public or pose a threat to national security; and

(2) whether a certificate of waiver, certificate of authorization, or airworthiness certification under section 44704 of title 49, United States Code, is required for the operation of unmanned aircraft systems identified under paragraph (1).

(c) REQUIREMENTS FOR SAFE OPERATION.—If the Secretary determines under this section that certain unmanned aircraft systems may operate safely in the national airspace system, the Secretary shall establish requirements for the safe operation of such aircraft systems in the national airspace system.

SEC. 334. PUBLIC UNMANNED AIRCRAFT SYSTEMS.

(a) GUIDANCE.—Not later than 270 days after the date of enactment of this Act, the Secretary of Transportation shall issue guidance regarding the operation of public unmanned aircraft systems to—

(1) expedite the issuance of a certificate of authorization process;

(2) provide for a collaborative process with public agencies to allow for an incremental expansion of access to the national airspace system as technology matures and the necessary safety analysis and data become available, and until standards are completed and technology issues are resolved;

(3) facilitate the capability of public agencies to develop and use test ranges, subject to operating restrictions required by the Federal Aviation Administration, to test and operate unmanned aircraft systems; and

(4) provide guidance on a public entity's responsibility when operating an unmanned aircraft without a civil air-worthiness certificate issued by the Administration.

(b) STANDARDS FOR OPERATION AND CERTIFICATION.—Not later than December 31, 2015, the Administrator shall develop and implement operational and certification requirements for the operation of public unmanned aircraft systems in the national airspace system.

(c) AGREEMENTS WITH GOVERNMENT AGENCIES.—

(1) IN GENERAL.—Not later than 90 days after the date of enactment of this Act, the Secretary shall enter into agreements with appropriate government agencies to simplify the process for issuing certificates of waiver or authorization with respect to applications seeking authorization to operate public unmanned aircraft systems in the national airspace system.

(2) CONTENTS.—The agreements shall—
   (A) with respect to an application described in paragraph (1)—
      (i) provide for an expedited review of the application;
      (ii) require a decision by the Administrator on approval or disapproval within 60 business days of the date of submission of the application; and
      (iii) allow for an expedited appeal if the application is disapproved;
   (B) allow for a one-time approval of similar operations carried out during a fixed period of time; and
   (C) allow a government public safety agency to operate unmanned aircraft weighing 4.4 pounds or less, if operated—
      (i) within the line of sight of the operator;
      (ii) less than 400 feet above the ground;
      (iii) during daylight conditions;
      (iv) within Class G airspace; and

(v) outside of 5 statute miles from any airport, heliport, seaplane base, spaceport, or other location with aviation activities.

SEC. 335. SAFETY STUDIES.

The Administrator of the Federal Aviation Administration shall carry out all safety studies necessary to support the integration of unmanned aircraft systems into the national airspace system.

SEC. 336. SPECIAL RULE FOR MODEL AIRCRAFT.

(a) IN GENERAL.—Notwithstanding any other provision of law relating to the incorporation of unmanned aircraft systems into Federal Aviation Administration plans and policies, including this subtitle, the Administrator of the Federal Aviation Administration may not promulgate any rule or regulation regarding a model aircraft, or an aircraft being developed as a model aircraft, if—

(1) the aircraft is flown strictly for hobby or recreational use;

(2) the aircraft is operated in accordance with a community-based set of safety guidelines and within the programming of a nationwide community-based organization;

(3) the aircraft is limited to not more than 55 pounds unless otherwise certified through a design, construction, inspection, flight test, and operational safety program administered by a community-based organization;

(4) the aircraft is operated in a manner that does not interfere with and gives way to any manned aircraft; and

(5) when flown within 5 miles of an airport, the operator of the aircraft provides the airport operator and the airport air traffic control tower (when an air traffic facility is located at the airport) with prior notice of the operation (model aircraft operators flying from a permanent location within 5 miles of an airport should establish a mutually-agreed upon operating procedure with the airport operator and the airport air traffic control tower (when an air traffic facility is located at the airport)).

(b) STATUTORY CONSTRUCTION.—Nothing in this section shall be construed to limit the authority of the Administrator to pursue enforcement action against persons operating model aircraft who endanger the safety of the national airspace system.

(c) MODEL AIRCRAFT DEFINED.—In this section, the term "model aircraft" means an unmanned aircraft that is—

(1) capable of sustained flight in the atmosphere;

(2) flown within visual line of sight of the person operating the aircraft; and

(3) flown for hobby or recreational purposes.